Tuscawilla

Stories of a Farm

F. JANE CARY

compiled and edited by Kaaren Cary Ford

A Blackwater Press book

First published in Great Britain and United States of America by Blackwater Press, LLC

Printed in the USA by Ingram and in the UK by Mixam

LCCN 2021949165

ISBN 978-1-7357747-1-8 (hardback)

Cover design by Eilidh Muldoon (www.eilidhmuldoodles.com)

Interior illustrations by Chloé Valerie Harmsworth (www.instagram.com/chloevalerienatureart.com & https://chloevalerienatureart.wordpress.com)

Blackwater Press
120 Capitol Street
Charleston, WV 25301
United States

www.blackwaterpress.com

F. Jane Cary
('Big Jane')

Tuscawilla Farm
Lewisburg, West Virginia

Cider

Contents

Foreword

This is a book of memories, stories of a large working farm from the 1930s to the early 1960s. The memories are those of Frances Jane Cary (1923-1995), my aunt. The farm is Tuscawilla Farm, just outside of Lewisburg, Greenbrier County, West Virginia. I daresay that there are not any farms like it now.

Tuscawilla dates to 1844 and encompasses 388 acres with six houses, one very large barn and, at one time, three other barns, four ponds, a large wood, and a formal flower garden.

The main part of the farm lies along Route 219 South. The Brick House can be seen at the top of the lane, along with the Cary House, the large barn, Mr. and Mrs. Edward Knight's house, and the chauffeur's house, where I grew up. The farm stretches down the back fields to the large woods, to Teaberry Road, across the road and up a hill where there was a "Red Dog" quarry. At this part of the farm there are two ponds and two more barns. Years ago, some of the Shorthorn cattle herd and the sheep were kept in these fields. The farm goes down Teaberry Road to the entrance to Whitcomb Road. Gene Hughes, who worked there for many years, lived near the sheep barn, while his brother lived in a small house along Teaberry Road on the back of the farm.

This farm had it all: a large apple orchard, ponds for fishing, horses to ride, a commercial dairy, prize-winning Shorthorn cattle and Tamworth hogs, turkeys, chickens, ducks, and sheep. It took a great many people to make a success of farming at Tuscawilla. It was its own little village where workers had to have diverse skills, from carpentry to gardening, to knowing how to work the various pieces of equipment, how to repair machinery, how to milk cows, and how to handle and care for the various animals. It was about as self-sufficient as it could get, and a wonderful place to grow up.

The Cary family has been a part of Tuscawilla since the early 1900s, when my grandfather, Frank Julian Cary, started work with Edward Wallace Knight. Through the years the descendants of these men continued to work together until my aunt's death in 1995. The Knight family still owns and lives on the farm, and although the Carys have left it, they still regard it as their home.

My aunt loved every inch of the farm; some would say its soil was in her blood. She thought it was important for others to know about this special place, not only because of its historic significance but also because this way of life is no more. She believed people should know about the experience and hard work of living on a large working farm, populated by many characters.

She taught many teenaged boys how to work, instilling in them a good, strong work ethic. Most summers, high school aged boys would work putting up hay, repairing fences, and working in the formal flower garden, which was her domain. Oh, how they hated the flower garden as they quickly learned that Aunt Jane could see the weeds they missed even with her back to them! As my brother Frank recalls, "she would sit on her wooden apple box with a small hoe, being sure that all the weeds were pulled. She could spot a weed from a mile away!"

Aunt Jane was a unique person. She knew just about everybody in the area, though she seldom left home. She was liked by all. Her interests included fishing, camping, gardening, cooking, keeping up with family and friends, sports, and history, especially Greenbrier County history.

My aunt was a loving person. She loved her family very much. She was full of hugs, laughter, guidance, and thoughtfulness. She loved to cook and often made large meals and invited friends and family to eat. She would also make meals to send to those she knew were not well. At Christmas she sent out at least a hundred Christmas cards, each with a hand-written note.

She was a loyal West Virginia University sports fan, and she followed the football games listening to Jack Fleming on the radio while watching several other football games on television and was able to keep track of each game and carry on a conversation.

After she stopped smoking cigarettes, she took up smoking a pipe as a way to wean herself off tobacco. When health difficulties arose, she continued her work on the farm by attaching her oxygen tank to a riding lawn mower. Having been born and raised at Tuscawilla and having lived there all her life, she knew what needed to be done and when it needed to be done and she wasn't going to stop working. She grew up in a farming family: her father, my grandfather; her older brother, Ed, my father; and her younger brother, Julian, had been the farm managers. She became the manager herself, one of the only women to work in that capacity, after the deaths of her two brothers.

The Cary family was extremely social: they always had company, friends and cousins stopping by. You could find many people there on Saturdays to watch the sports games or bet on horse races or play card games like Set Back. Often these folks would stay for supper. Aunt

Jane seemed to always have enough food on hand for any planned or unplanned meal. I think she could have fed General Patton's army if they happened to stop by!

Aunt Jane was a good cook, especially with any cut of meat. Her secret was cooking it slow. There was always an assortment of condiments to go with whatever the main course was: pickles, olives, and her favorite, fresh cucumbers and onions in vinegar ice water.

Her fresh oyster feasts were well known. In the winter months, friends and relatives, along with her, would order gallons of fresh oysters from RIPCO (Ronceverte Ice and Produce Company). On the day of the feast, Aunt Jane would start early getting things ready for dinner. Side dishes were prepared and then, closer to the dinner hour, she would make the oyster stew, which was made with cream and butter. Then, when folks started to arrive, she would start with slowly frying the oysters in cracker crumbs, for a meal of oyster stew and fried oysters.

She did occasionally have a disaster. Once she decided to cook a whole duck on the grill. I don't think she'd ever eaten duck, much less cooked one. One thing led to another, and before she knew it, the grease had caught on fire and flames were reaching the second story windows, burning the side of the house. She never tried to cook duck again!

Frequently, especially in the summer, there would be impromptu family picnics in her yard. Word would quickly spread from cousin to cousin and the gathering turned into a family reunion. There was always good food at these picnics and lots out of door games to play.

Once she decided to have a luncheon for some older cousins who were in town visiting. When I wandered in, these ladies were laughing and giggling, telling stories and such. I had never seen them so happy and so funny. I asked for a glass of lemonade, but Aunt Jane wouldn't let me have any. It seems, unknown to the old ladies, the lemonade they were drinking had been spiked. Years later they would talk about what fun they had that day.

Aunt Jane was very social in high school and was generally a class officer and was active in the school social clubs. One meeting of a young women's club was held at Betty Sydenstricker Woodward's house, Montwell. Here was Aunt Jane in her homemade feed sack dress eating caviar that Betty had bought at Houck's store. She really thought she was in high society, and I wish she'd written down that story. She didn't graduate from high school as she spent seven months in the hospital with rheumatic fever. She had wanted to be a teacher, and I know all who knew her and worked with her would agree that she was an excellent teacher, just not in a classroom.

She liked to fish, and quite often on a weekend would load us up into the old Jeep station wagon and head to one of the ponds on the main part of the farm. She had an array of fishing tackle and other fishing gear. Sometimes she would join others at the Greenbrier River to fish and then camp for the evening. Once she and Jackie Goodall decided to fish on the other side of the Fort Spring Tunnel. It was a cloudy day, and the tunnel was very dark. Jackie told Aunt Jane to stay put while he walked the tunnel making sure it was sound. When he stopped to check the tile, he turned and bumped into Aunt Jane, who was silently walking behind him; she wasn't about to wait!

Aunt Jane was the sort of person who simply enjoyed life. She enjoyed her family, her friends, and the farm. She enjoyed the out of doors, whether it was driving in the country, or through the woods at the farm, or camping.

She was a great letter writer, and kept in touch with friends and family by writing long letters on legal tablets. To get one of her letters was a treat, as she always included what was happening on the farm as well as what was going on in the area. She'd send cards to those hospitalized or home bound, and often at Christmas she would send one of her homemade flower arrangements. She would always make certain flowers were on the graves of family for Easter and Christmas. When she noticed that her grandfather Dick Hayes' gravestone had deteriorated to the point that it could no longer be read, she got in touch with all the Hayes cousins to raise funds for a new one, which was installed in the Hayes family plot at the Old Stone Church Cemetery.

When we were home sick from school she would come to our house – just on the other side of the barn from the Cary house – loaded with coloring books and crayons, and we would spend hours coloring; what fun we had! We almost looked forward to being sick.

One time she and my mother decided to cut my hair, each working from a different side. I ended up with my bangs at the hair line. When my dad came home and saw what they had done, he took me to Mr. Ferrell's barber shop and the rest of the summer I had a boy's haircut, which in 1956 was not something girls had.

Aunt Jane was very interested in Greenbrier County history, and she kept scrapbooks of articles from the local newspapers. Folks knew of her interest and her scrapbooking and would send her photos, articles, and such. Quite often those visiting her would spend their time looking at the scrapbooks. At her death there were thirty scrapbooks, some composed of family photos and one of clippings from weddings, but most were articles and photos of individuals she knew, and articles of local and state history. These have been donated in her memory to the Greenbrier County Historical Society.

Sometimes in the winter, especially when there was a large snow on the ground, Aunt Jane would quickly organize a sleigh riding party. Cousins and friends would show up with their sleds, a bonfire was lit in one of the fields, and a massive tub of hot chocolate was made. We kids would have a blast sledding down the lane trying not to run into the thorn hedge. The only bad part was walking back up the lane dragging your sleigh! The walk back was never fun. One time the sleigh riding party was held on the back field, where if you went too fast you would end up hitting the fence around one of the ponds. If the sleigh hit the cow path that went around the hill near the barn, you could end up flying off the sleigh into the snow. While we did have plenty of accidents, no one was ever hurt and it never affected our fun.

Christmas was a time when Aunt Jane's house was fully decorated, inside and out. The enclosed front porch was where she kept the Christmas tree with gifts underneath. On the bookcase along the porch wall she would make a scene out of moss gathered in the farm woods. There would be a mirror as a pond, small cardboard houses and a church were arranged along with small animals. The mantel above the living room fireplace was covered with ground pine and then decorated in red. There were candles and floral arrangements throughout the house. Christmas Eve was a time for a large family dinner. Everyone would come together for a meal full of everything she loved to make. As years went by it was more of an open house, with folks coming and going, rather than a sit-down dinner.

One of the best stories is when Aunt Jane sent my brother, Frank, and our cousin, Billy Mike Landers, to dig thistle in the field near the woods. It was a hot afternoon, and these guys took their time walking to the field, plus digging thistle is actually hard work. After digging up two or so thistle, Frank and Billy Mike sat down under a tree to rest. Before too long both were fast asleep. They woke up just before 5 o'clock and dragged themselves back to the barn, acting like they were very tired from their work. They attempted to tell Aunt Jane how hard they had worked, and how many thistles had been dug, but she knew they had been asleep, so their story didn't sell. She sent them back the next day to finish the job, and they learned they couldn't get away with anything. It took years for them to find out how she knew they had goofed off!

The Carys and the farm were always actively involved in the West Virginia State Fair. The most prize-winning bull of the Shorthorn herd was solid white. We called him Whitey. He was very gentle and would let us rub his face and give him hugs. Once at the State Fair, where competition was especially stiff, someone decided to wash him with bluing so he'd look his best. Imagine their surprise when the bluing didn't make him whiter and brighter, but instead turned him blue. I always felt sorry for Whitey as he had to endure several baths that afternoon and evening to remove the blue before the competition.

I would like to mention three men who worked at the farm while I lived there, who aren't featured prominently in my aunt's recollections. Charlie Hughes was a person who could do just about anything. He plowed the fields, fixed equipment, and fed and cared for the hogs, among many other tasks. Charlie's main job was feeding and caring for the Tamworth hogs. He would always warn us to get up in the door frame of the apple room when the mother hogs and their litters were being moved from the barn to the field because mother hogs can be extremely protective of their piglets and become aggressive. He could build just about anything, from a bridge to a hog house. Charlie lived in a small white house on the back side of the farm, next to Teaberry Road. He was always ready to help us with our 4-H projects and never tired of our questions.

Another person who was always part of the farm was Charlie's brother, Gene. Gene lived on the old Anderson place on Teaberry Road across from the farm. He was the first person to get to work every day, and was generally the last one to leave, as he was the primary person responsible for the milking. Every morning he would walk down his hill, cross Teaberry Road, climb the fence, walk through the farm's woods, through the field and up the hill to the barn. Often the cows would walk with him and sometimes they were waiting outside the barn. Once the milking was done, and the milk parlor and the milk storage area were cleaned, he would join the men in other jobs: putting up hay, driving tractors, fixing equipment. In the afternoon he would do the second round of milking and walk home. These were long days for Gene because one thing farmers know, especially dairy farmers, is that livestock don't recognize Daylight Savings Time! Dairy cows need to be milked the same time, twice a day, seven days a week.

David "Bud" Williams was a mainstay of the farm. He started working there as a young man and didn't stop until he was quite old. Bud was a person who just about everyone in town knew and respected. He would drive to work from the top of River Road in Ronceverte every day no matter the weather. He would drive to town every morning to get the mail, take Miss Elizabeth Knight to her appointments, and drive the produce to the train station to be shipped to Charleston. As a preschooler, I would ride with him to town every day. He also did odd jobs on the farm, and taught many of us, myself included, to drive. I remember when men would stand on the street corners of downtown Lewisburg to see who was in town and to visit with folks as they walked by. Quite often this would happen to us when I would ride with Bud to get the mail or pick up groceries. Everyone in town seemed to know Bud and would ask him to stop to talk, to find out how things were at the farm and how deep the Greenbrier River looked when he came to work, and things like that. Imagine my mom's response when someone called her when I was in high school to report that I had been seen riding around town with a black man. After she laughed, she said, "Where have you been? Kaaren has been riding with Bud since she was three years old."

The views from Tuscawilla are spectacular. Looking east is the gap, where Kate's Mountain sits in between Greenbrier Mountain and White Oak Mountain. Looking west is the mighty Muddy Creek Mountain. How could any view be better than these? Growing up there with the smell of the apple trees in bloom, the sounds and sights of the farm animals, the beautiful flower garden, and the green fields with the mountains in the background, was what I imagined heaven to look like. With this in mind, those at Blackwater Press have created a webpage to accompany this book, full of pictures from Aunt Jane's time on the farm, as well as those that show you what it it looks like today. Just head to: www.blackwaterpress.com/tuscawilla.

I hope that by sharing her stories, others will have a better understanding of what farm life was like during the 1930s through to the early 1960s, as well as a better appreciation for the work that farmers do. I also hope that the readers may understand what a wonderful place Tuscawilla Farm was and is. Having grown up there myself I know so.

Editorial Policy

Transcribing from handwriting, especially from yellow paper that has yellowed with age, is hard work. In instances where words were illegible, the best guess has been made for sense and clarity. Punctuation and spellings, especially of names of animals, has been made consistent. Spellings such as "setting" for sitting have been retained. All of the men in the Knight family are called Edward; this can be confusing for the reader who doesn't know the family. Aunt Jane tended use Mr. E. W. Knight and Edward Wallace and Ed. W. to refer to Edward Wallace Knight. I have tried to make this consistently Edward Wallace. Ed Knight or Mr. Knight refers to his son, and Eddie refers to the Mr. Knight of my childhood. There are undoubtedly places remaining that will remain confusing, especially after my father George Edward (always referred to as Ed or sometimes Brother Ed) enters the mix. The family trees should help to alleviate any lingering confusion. Aunt Jane wrote much of what follows over a long period of time, as individual stories not intended for publication, so there are slight repetitions. I have attempted to leave it as untouched as possible, but have made some minor edits for clarity.

And inevitably, some of my aunt's language has not aged well; it is a product of her time, location, and culture, though by today's standards some of her references to African Americans would rightly be considered offensive, she truly was progressive for her time.

Acknowledgements

Aunt Jane's scrapbooks of family history and Greenbrier County history are housed in the North House Museum, Lewisburg, West Virginia. This book was transcribed from her piles of handwritten stories on legal tablets. Some of these stories were excerpted in *The Journal of the Greenbrier Historical Society*, vol. X, no. 2. 2019, and this edition represents the complete collection of my aunt's writings about this lost way of life.

Many people have helped me with this edition by offering suggestions, definitions, or providing the practical assistance of transcription and editing, and I would like to thank in particular the Knight family: Dana Knight Henderson, Edward D. Knight III, Mrs. Edward D. Knight, Jr., Whitley Knight; Debra Hutcheson, and Frank Cary. I also thank Eilidh Muldoon for her beautiful rendition of the farm, based on our cider labels and an aerial photograph.

Kaaren Cary Ford

Knight Family Tree

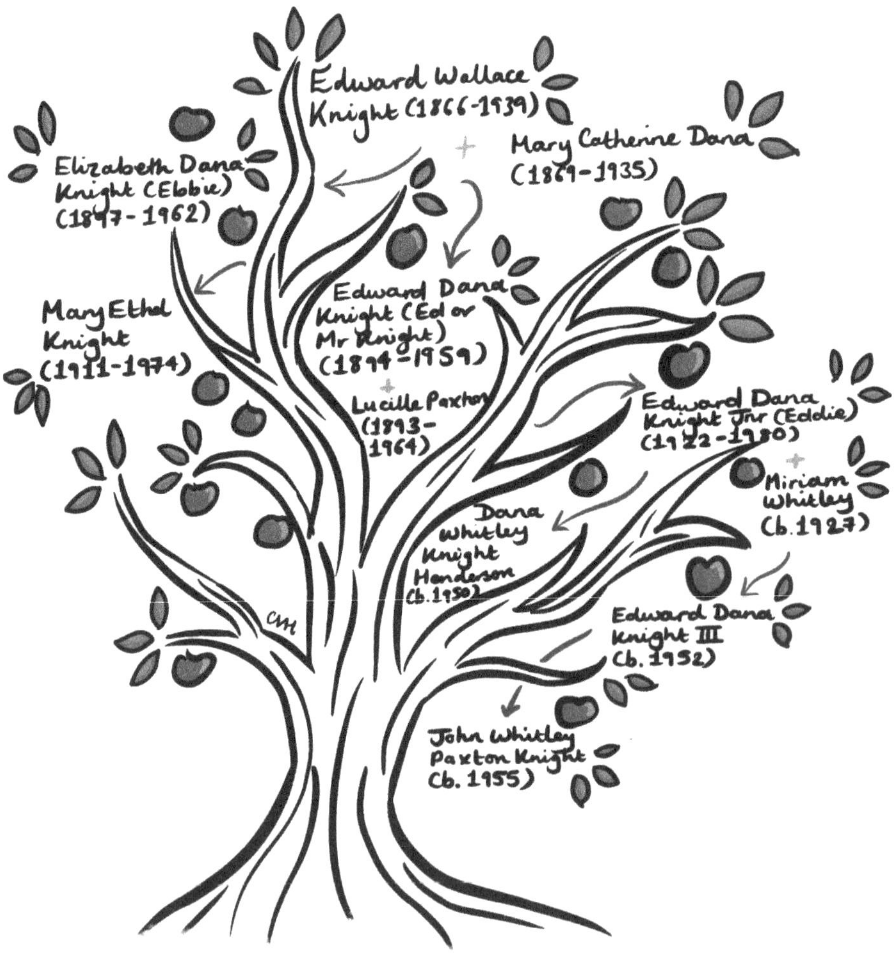

Cary Family tree

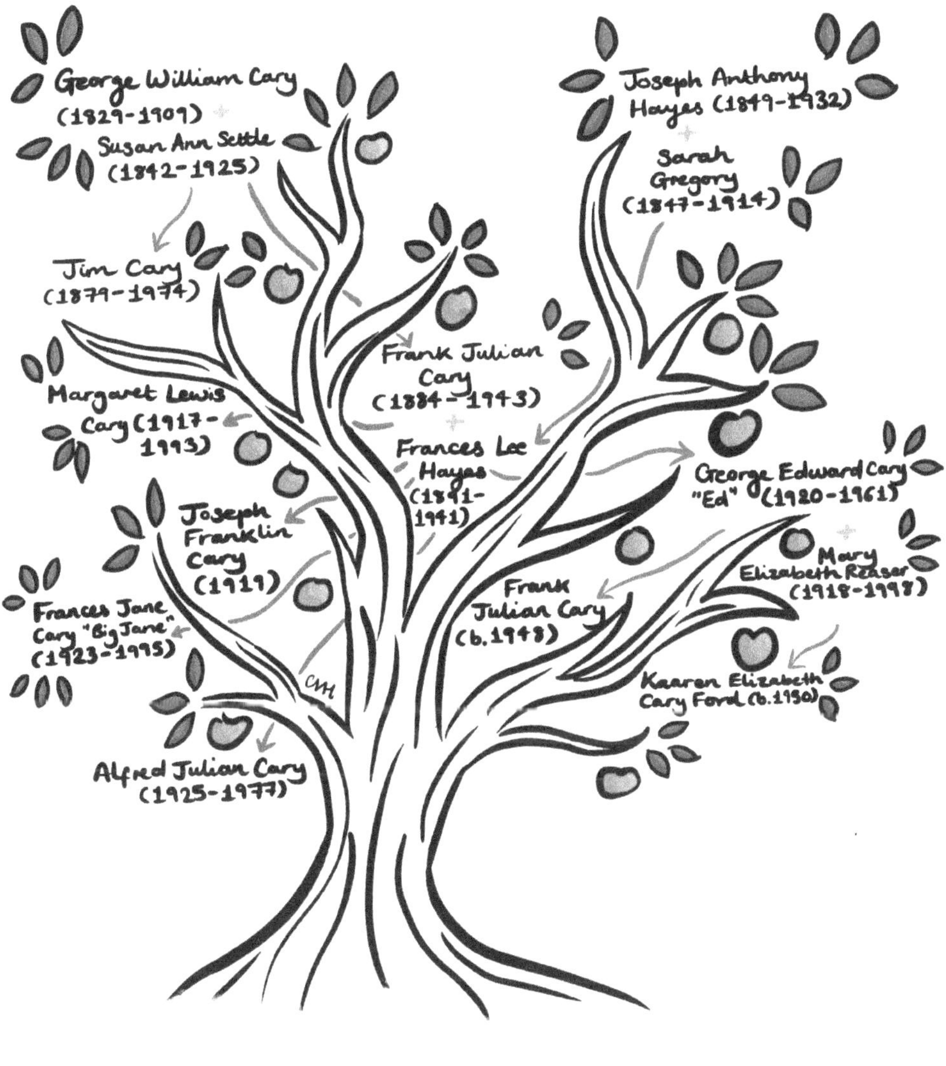

History of Tuscawilla Farm

Thomas Creigh (1766-1847) was a man of prominence, wealth, and esteem. He came to this country from Ireland in 1796. He owned most of the land north east of Lewisburg, then in Virginia, halfway to Frankford, which was the first permanent settlement in Greenbrier County, and south of Lewisburg, reaching to the Greenbrier River. He married Margaret Williams on August 1, 1801. She was the niece of Colonel John Stuart, the father of Greenbrier County, a surveyor, Indian fighter, and chronicler of pioneer history of southern West Virginia. They had nine children: their daughter Jeanette McCreary Creigh married Reverend David Preston, a Presbyterian minister on August 12, 1840. I was told that they served as missionaries and ministered to the people of the western mountains of Virginia. Then hearing of the need among the southern Indians, they moved to Florida to serve the Seminole Indians. After a few years Reverend Preston came down with swamp fever and was forced to return to the mountains. They bought the land now known as Tuscawilla Farm from her father, Thomas Creigh. They called their farm a musical name, Tuscawilla, a Seminole Indian word, meaning two lakes. There were two large ponds on the farm.

Reverend Preston started building their house in 1844. Slaves were used to make the brick on site. In those days, everyone tried to make a better brick than the next man. The Brick House at Tuscawilla appears to be a two-story house, though it is really four stories consisting of a basement, first floor, second floor, and a large attic. The front of the house has nine windows with small panes. The third floor, the attic, has two very large rooms. Each room has a small window with very small panes. When in the attic, one can see large floorboards from all kinds of trees. The house has four very large chimneys with a fireplace in every room. The upstairs has four bedrooms with a main central hall. The main floor had a large entrance hall, two bedrooms, a large parlor, a dining room and a kitchen with a fireplace. There are two rooms over the original kitchen, which is currently the dining room. Several years ago, on the south side of the house, a new kitchen and pantry were added. The front door measures forty-five inches; it has panels of frosted glass and the door is put together with wooden pegs. It has a brass door knocker and brass locks. Wide boards of pine and oak were used all over the house. This house was one of the finest of houses of all times. In the back of the house there was a large well and the slave cabins. The yard was full of many large trees.

Before Reverend Preston's death, his swamp fever returned. He became a man of rages; at different times he would beat his slaves. He became enraged with two slaves over stealing meat. Mrs. Preston knew it was the fever not the man. A year or so later the fever returned, this time his rage was more than he could bear. He called two male slaves upstairs over the kitchen, which is now the dining room, to what was his study. He shot and killed one slave on the spot. He shot the other slave trying to run away. He died also. There was a blood spot on the floor for years. Ethel Knight took everyone to see this spot. The Knights had colored maids; they would not sleep upstairs after the two rooms were turned into bedrooms. Elizabeth Knight, known as Aunt Ebbie, had a great maid, Myrtle, who used the room in the daytime to change her clothes and rest. She also did the ironing up there, but she never did go up to that room at night as the ghost of the slave was said to walk.

Tuscawilla saw the army of the South and the army of the North passing by on the North-South Turnpike. The residents saw smoke, heard the battle cannons and all during the Battle of Lewisburg. The beloved Southern fighting men, returning home in rags, their heads hung low in defeat, found Tuscawilla unharmed! The Prestons blocked the front lane with pine brush and showed no lights in the windows at night. Unharmed, Tuscawilla stood, but it was touched in many other ways. Three young men went to war, the finest horses went, as well as other livestock. Jeanette Creigh Preston saw her brother's family meet with such sorrow during the war. While her brother, David Creigh was away, a Yankee looter stopped and went into the house, upstairs into a bedroom, where Mrs. Creigh was setting with their sick daughter. When David Creigh returned home to find a Yankee in his daughter's room, he shot and killed him, then hid the body and rode off to find friends and neighbors to see what he should do. They all decided to get the body hid for good.

David Creigh went home, and with no one's help but two slaves, dropped the body into the well. Now, I was told that one of the slaves was a troublemaker, and he ran off that night and told someone who got word to the Yankee Army, which was west of Lewisburg. The Yankees came for Mr. Creigh and took him, and his wife and daughter, to the mill on Sinking Creek where a military trial was held. No one was given time to speak out for him. His wife and daughter were made to walk back to their house. David was taken to Virginia, where he was hanged on November 7, 1863, at Brownsburg.

Tuscawilla Farm was empty for a few years. Eventually, Mr. and Mrs. George Bloomer from Ohio bought the place. They lived on the farm on and off between 1863 and 1886. Mrs. Caroline Bloomer was a lady of wealth: she farmed and maintained the finest stable of horses. George Bloomer was older than Mrs. Bloomer. He was Vice President of St. Lawrence Lumber Company which had just opened a very large mill on the Greenbrier River at Ronceverte. The home office and biggest mill were in Ohio. Mrs. Bloomer bought two other farms, both owned by Jeanette Creigh Preston's family. Dr. Thomas Creigh, her brother, had a farm south of Tuscawilla that was called Tallassee. His farm was as large as Tuscawilla and the house was a two story log one. I never saw such big logs! In later years the logs were plastered over. It was a great country home. I was always told there was a tunnel from the dirt basement up back of the house to a small underground room that had an escape door. This was built as an escape route when the Indians raided.

The first cabin on that farm was only two rooms; a bigger cabin was built later. In 1985, the State Fair bought the house and the lot and then tore down the house and cabins in 1986.

Mrs. Bloomer also bought a third farm; it was the David Creigh farm called Montescena which had been empty for a long time.

Caroline Bloomer was the belle of the springs. She had a great social life in and around Hot Springs, White Sulphur Springs, Salt Sulphur Springs, and Sweet Springs. This was around 1884. She was noted for the finest carriage horses in and around all the Springs.

Tuscawilla was play land for all the young southern belles and gay blades. George Bloomer was seldom home; he spent most of his time at the home office or buying timber. He was an older man and didn't care for the gay life. He and his wife drifted apart, and she started a romance with her coachman whose name was Simmons. This was a great county scandal. They had a daughter. Caroline went to D.C. to have the baby and when she returned, she moved the family into the house at Tallassee. She and Simmons raised the daughter, who was sent to schools and had the best of everything.

George Bloomer left Caroline. He returned to Ohio, as most of the timber had been cut around Pocahontas and Greenbrier Counties. Caroline stayed at Tuscawilla but spent the winters in D.C. where she died.

Tuscawilla was empty once again for a few years. It was sold to a Joseph G. Robertson, who was a sawmill man. He built sawmills on the farm, and farmed. One mill was in the bottom above the old log cabin on the back road. The other was built on the right side of the woods in the flat area. He had a large sawmill at Frankford. He cut pine timber here on Tuscawilla.

I knew Mr. and Mrs. Robertson's daughter, Ethel Robertson Stone. She had a great thing going for Tuscawilla Farm. She loved it. She was old enough when they came to the farm to remember how hard it was to get the horse and wagon up the lane. I remember she always said the Bloomers planted the Washington Thorn hedge along the lane. It had grown together, which made it very difficult to get the horses to travel up the lane.

They found the house run down. Mr. Robertson paid around seven thousand dollars for about two hundred and sixty-five acres of Tuscawilla. The house still held lots of the Bloomer's furniture and fixtures. Ethel still had a chest and some lamps left by the Bloomers. They only lived there two years. Ethel Stone died in 1975. She always told the story about the slaves being killed.

I never was so sure about the very early part of the Knights' purchase of Tuscawilla. I was always told that Colonel Eugene Dana and his wife, Maria Swift Dana, bought this farm for their family. A few years later their son-in-law, Edward Wallace Knight, bought part of the farm where the house Tuscawilla was sitting. Both families used it in the summertime. Mr. E. W. Knight wanted a summer home for his family so Tuscawilla was the place. It started out as a family farm: a couple of milk cows, a few sheep, hogs here and there, chickens, turkeys, a vegetable garden, lots of buggy horses, work teams, a couple of riding horses.

Edward Wallace Knight married Mary Catherine Dana. They bought the farm in 1909. They paid two thousand dollars for it. Mr. Knight wanted a place in the country not too far from the railroad. He was the General Counsel of the Virginia Railroad Company. He needed to be near the railroad; he traveled most of the time on the K. and D. with Col. Dana. He shared the farm for a time. Mrs. Dana died a year or so later and Col. Dana married a red-haired dancer from New York. About half the time when the Knights came to stay at the farm, their stepmother Dana had run off the help. She tired of the Springs and missed New York, so they moved back before long.

Mr. Edward Wallace Knight and his wife, Mary Catherine Dana Knight, and their children Edward Dana, known as Ed, their only son; their daughter, Elizabeth Swift and then fourteen years later, daughter Mary Ethel, lived many years here on Tuscawilla. Mrs. Knight spent a long time beautifying the house, the grounds, and her beloved flower garden. Her husband also spent many happy years planning his farm, planting, developing his orchard, and raising hunting dogs. I was always told it was a place of quiet and easy resting for Mr. and Mrs. E. W. Knight.

The beautiful brick house is reached today as it was in the years gone by, by a long thorn hedge planted as I said by the Bloomers. This hedge encloses the lane, which was dirt. The yard is full of oaks, red and yellow maples, tall swaying hemlocks, a big elm, and dogwoods. The north side of the house is flanked by an enclosed flower garden and large white barn. The house had a large front porch with steps in the middle.

The porch had six white posts and a railing. There were big hemlocks on each side of the steps. The house also had a door on each side of the house with steps.

Mary Catherine Knight died in September 1935, and Edward Knight died in 1939. They spent only a short three months on the farm a year; the rest of the time they lived in Charleston. Mr. Knight was a great sportsman; he loved dogs and horses, played golf at the Greenbrier every day he was at the farm. They cherished all who worked for them; they were both caring for all others. Mrs. Knight most sincerely, truly, shared in her part of Tuscawilla with an iron hand. She was respected by every person in Greenbrier County. When she came out every day to the flower garden every hat came off. The younger folk cared for her. Tuscawilla was a grand party farm: they entertained the family and friends several times every summer. Mrs. Knight had more flowers in her flower garden than anyone could ever know. She also raised a cut flower garden full of roses, delphiniums, stock, canterbury bells and cockscomb, out behind the brick house. She put flowers in the Old Stone Church every Sunday she was here. She carried flowers to all the ill she knew and funeral flowers for any of the farm workers' families.

The cook was Rachel Deaver, who was a very young unmarried gal from Teaberry. She stayed in the cook's house when the family was here and moved back to her home in the early years when the Knights were not at the farm. Later they helped her build a small house in Fairlea. From the first I can remember, every Sunday morning, rain or shine, after breakfast there was a horse and buggy tied at the back door. Rachel drove home and to church in Fairlea every Sunday. She was the last person around here to drive the horse and buggy. The man who milked on Sunday morning and evening harnessed it up for her. The old horse was named Princess. The milkman had to unharness the horse and put her away for the evening when he finished his milking. She turned into a cute little old lady, chubby one, always had a small white dog, which was her life. Rachel was terrified of storms; she would go hide.

Back then the brick house had a root cellar, a woodshed, a smokehouse, a chicken pen around the back along with a well and cistern. The horse lot was near the flower garden gate. The cow and calf lot was at the upper end of the flower garden. The small red barn stood in the far corner of the flower garden. The well was at the front door of the chauffeur's house.

Late in 1917, the Knights started to redo the brick house, to build the barn and the cook's house, the Cary house, and the garage apartment. The First World War slowed things down for a year or two. Most of the big timbers in the barn were cut in Pocahontas County, West Virginia. Some of the lumber was cut on the farm but not much as when Colonel Dana got the farm it had just been timbered by Mr. Roberston.

Grandpa and Grandma Knight had big family gatherings, played kick-the-can in the barn once a week, and most all summer long. On the fourth of July they had their biggest parties. Big Ed – E. D. Knight – put firecrackers off, and they were bigger and bigger each year. Homemade ice cream was served, and cake, and the servings were big. Most of them rode to the Greenbrier River behind the farm to swim! Six or seven riding horses, three or four working horses, or mules, would be taken. Great-grandpa, Mr. Edward Wallace Knight, had one party I remember. There was an Eastern United States Educational Meeting of some sort at the Greenbrier. Man, did this place shape up! Everything shined. Twelve men worked ten hours a day to get ready.

A big friend of his was the president of Dartmouth. He and his wife were here. The president of Harvard was here, as were the presidents of Cornell, Yale and Princeton. Mr. Edward Wallace Knight was such a proud man, he gave all men a big walking stick, to all his guests. They didn't miss a thing on this farm. Edward W. Knight and my dad got along so well; he kept Dad on when he bought the place from Mr. Cartmill. Anyway, Dad talked him into a matched white pair of Percheron horses – mare not purebreds – just common farm horses, bred them and out came a solid white mule called Kaiser, and a red mule called Arbury, both born practically the same day, and great- grandfather Knight was the proudest person on earth over his team of mules. My daddy was equally as proud as he was. The mules lived their entire life here on the farm.

E. W. Knight started the purebred hogs. He brought his first hog from Canada in 1923 or 1924. There were hogs here before and they were called Sand Brier, and there were half wild hogs from Indiana. They were red, and very dangerous. We couldn't walk in the orchard sometimes when they were out. They would fight anybody, or anything, over their babies. He got rid of those and put in the purebred Tamworth hogs.

E. W. Knight was great. He also put in purebred Jersey Cows. He had eight at all times. He had two bulls that I remember well. He named one of them John Dickinson and one of them Charlie, and one was the blackest, meanest bull there was.

Tuscawilla Farm has had many folks, so many in my life. I was born here in this same house in 1923, as Ed was in 1920, as Julian was in 1925. My memories come along around 1928. By this time Tuscawilla Farm was a fast-moving farm. The brick house had been redone for the second time, the barns were all up and going full speed, they were building Mr. Ed Dana Knight and Mrs. Lucille's summer home on the hill. There was a fifty-acre apple orchard, the flower garden had been laid out and planted by a Mrs. Bell from Cincinnati, Ohio. Fourteen or more horses, jersey cows, a team of mules, and by this time the big Tamworth hog herd was going full blast. The brick house by then had a full time cook, a maid, a houseman, a gardener, and a chauffeur.

In the flower garden there were two apricot trees, the best in the world. By the garage apartment there were four big cherry trees. Alongside of the flower garden by the road were four more cherries. In the chicken lot we had plum trees, and a couple of peach trees. Lynn Green, one of the men who worked here, took care of them all. Mrs. Knight had a cut flower garden by Rachel's little house. On one side next to the house, she had her garden, also horseradish, rhubarb and a big mint bed. By this time Miss Elizabeth had a one- room house behind the cook's house where she painted.

Back then, Mrs. Knight had a lamb killed once a week, on Thursday afternoon for the weekend. Dave Deaver was in charge of killing the lambs. He would pick out one Thursday morning, put it in a certain place in the barn. At four o'clock Thursday afternoon he would get the lamb, bring it outside the hog barn under an apple tree, and I would just love to take the city cousins down to watch this. They'd all get sick, go home crying and mad at me forever, but I always liked to tease them with it. But, anyway, Dave killed the lamb, dressed it under the tree, took the hide off, salted the hide, stored the hide, cleaned the lamb, took it over and put it in the basement of the brick house. We didn't have a freezer room or coolers. The next morning, he would go back over and cut it into leg of lambs, front quarters, and chops. Grandmother Knight always had lamb for the weekend, always, when she was here. We never did ship lamb to Charleston, only had lamb during the early summer and late, up until late fall. We also got some lamb every week, along with chicken. Jim Early, the old colored fellow, would kill each family three or four frying chickens or old hens also, and a duck or two once in a while.

Lynn Green always buried cabbage, potatoes, turnips, hanovers and one hill of apples. At this time when the Knights went back to Charleston for the winter, they put three quarts of Jersey milk and one quart of cream in boxes twice a week to send to them. David Williams drove these boxes to Ronceverte to the train station, where they were put on train number thirteen in Ronceverte at fifteen minutes to twelve. Once or twice a week three dozen eggs were put in a box with three pounds of butter and also sent to Charleston.

Mr. E. W. Knight was at the farm for a day every two weeks or so to bank meetings. On these trips he would take home pork, ham, apples, and vegetables out of the hills, chickens, ducks, and turkeys. A couple of times a winter Dad would go down with things they needed, mostly moonshine.

Oh! Back to the brick house. I left out one main person. She was Ethel's maid, Nora, a tiny little German lady from Cleveland, Ohio. She was with Ethel forty years. She saw to Ethel's food, clothes, laundry, and her behavior at all times. She had an iron hand. We Cary kids were her pals. She read to us and played games and sang.

For twenty-five years the farm had two station wagons, just alike, one for the brick house, and one for Dad, plus two Buicks, two Fords, two Pontiacs, and one Jeep. Mr. E. W. Knight was a Packard man.

The Knights had kin folk up in Greenbrier County, so Tuscawilla was the place they all gathered. Mrs. Knight's brother, John Dana, had his home on the hill between the farm and Lewisburg. It was only two acres and a lane. For years it was their summer home. Later on, Dad had to look after their place when they were gone. He cut grass and cared for the twelve apple trees they had.

On over the hill, backing part of Lewisburg, were the Smiths. Mrs. Knight's sister Elizabeth Adelaide Dana married Isaac Noyes Smith. There was a cabin and two hundred acres over there. Tuscawilla farmed that place, looked after the cabin, wintered part of the horses there, cut hay, planted the garden, helped care for a tree nursery. They were here only in the summer. Tuscawilla parties were big for these folks. Lots of times there would be Smiths, Carys, Danas, Knights, Currys. They always had cocktail parties once a week. All the children played at the barn, loads of kiddies.

By this time, in the thirties, the flower garden had been changed. Mrs. Bell was here again for ten days, and the sundial and flag stone walk in the center of the garden had been taken out, so it was all grass in the center. All the beds had new dirt and the flowers were reset and new flowers planted. The orchard was going strong.

Aunt Ebbie was in Europe studying painting. Ethel was in school, young Eddie, E. D. Knight Jr., was in school in Charleston. Mr. Ed was doing a lot of hunting and had been to Kansas to a reunion for his Calvary outfit from the First World War. I think Mr. and Mrs. Ed and Elizabeth went to South America in the late thirties.

It was nothing for Dad to go to the barn at six in the morning to find three or four fellows who hadn't been to bed and had had a snoot full, like Pat Koontz, Arthur Hill, Sid Davis. They went to milk or fed the bulls, you talk about happy-go-lucky fellows, they were!

Mr. E. W. Knight had Jersey bulls, just one at the time. He named them after his friends, so when folks came to visit or stay, they all had to go see the bulls. Mrs. Knight put the ladies up over the kitchen and the men folks in bedrooms.

It was in the late thirties when Aunt Ebbie got a couple of horses to show. She showed in a couple small shows. She had a big horse, a sorrel call Highball. He was a colt out of a fancy mare here bred to a fine horse in Warm Springs. This colt all but whipped everyone on this hill to break. They tried all winter. He pitched everyone off, someone would come every week to see if they could manage. That spring when they plowed the corn ground, they put a saddle on him, tied three hundred pounds of oats on the saddle, put a horse and rider on each side and took Highball to the corn ground and worked him up and back for an hour every day. When the corn ground was about to plant the colt could be ridden. He was one devil high.

A young lad from Charleston by the name of Marshall Bond – he was one good-looking man – rode Highball for four or five seasons in horse shows, mostly in the summer. They went to small shows in West Virginia, Virginia, and Ohio, and did well if the devil horse wanted to. Aunt Ebbie showed a dark chestnut brownish-black horse call Barronwood. Before that Marshall showed the brother of Barnwood, who belonged to my dad.

The first house man I can remember was a darkie, man by the name of Harry Curry. He lived where Herb Montgomery built his house on the Teaberry Road. Harry helped to milk when the rest of the men were doing crop work. Harry lived to nip. His wife, Laura, did the washing and ironing for the brick house. Laura was a true lady. The gardener was Dave Deaver. He was Rachel's brother; he lived in Teaberry. Teaberry is a bluff overlooking the Greenbrier River, a lovely place. He walked twelve months a year to work.

The first chauffeur was Joe. I can't recall his last name, but it started with a B. Then came Marion Cooley. He was their full-time chauffeur; he was from Lewisburg. Marion was the first man to live in the garage apartment part-time. He would stay up town with his family the rest of the time, which was not to Mr. Knight's liking. Marion was bad to nip, so Mr. Knight wanted him here. At that time, he had the Packard, a Ford coupe, and the house station wagon to care for.

Oh! Back to Dave: he cut the grass with a gang mower pulled by a horse, then and a couple of men would cut it by hand with push mowers.

Another thing that was done early Sunday morning, every Sunday, was the house man had to turn off a gallon of homemade ice cream Rachel had made earlier, crank by hand, then set aside to season.

By this time, we're talking forties now, there was a huge grape arbor planted just below the flower garden, and as best I can recall there was twelve different kinds of grapes. You could smell them a mile away when they would ripen. There was always three beds of strawberries, one to be plowed under that year, one eating out of, and one new one. Blackberries, red raspberries, gooseberries, and currants too. Mr. Bud Legg and Lynn Green took care of these.

The vegetable garden was big in those days. Bud Legg and Lynn Green were the gardeners. It was two and a half acres. Lynn had three big hot beds behind the garage apartment, also there was a large tree nursery. The garden was started real early each year. They always had everything: big onion bed half an acre, rows of leeks, one hundred fifty staked tomatoes, rows of pole beans. They had a big row of celery. For years, every spring they would pick up a load of rocks and haul them away. The celery was buried in the row; this is how you bleached it.

One year Ethel wrecked Aunt Ebbie's new Ford Roadster. She was learning to drive and left the big yard, plowed into the hedge, tore it up but good. By this time Mrs. Lucille Paxton Knight had her mother living with them. She had a full-time nurse with her here and in Charleston, Lena Arbaugh. Lena's dad was living in Tenderhouse, the cabin which had been re-done with rooms added. Grandy Paxton was one real frail lady, so cute. We did a lot of playing on the hill with Eddie, Edward Dana Knight Jr. There were so many rules especially about when we could play.

By this time the railroad, which ran up the front of the farm to Lewisburg from Ronceverte, had gone broke so the right of way came back to the farm. It was a big job, took all summer. They used horses and pond scoops to put the ground back to crop land, built fences, and lo and behold next year Mr. Knight decided to plant the hedge up the front! Yes sir, all the front we had was planted all winter. He found a nursery with the hedge and ordered one thousand. They took a horse plow and plowed a trench, hauled manure, put in the trench, set out the hedge, hauled water all summer on a sled in barrels. It grew everywhere. In a short time, a few years later, they moved Route 219 over to widen it. Down came the fence, up came the hedges. Thank goodness.

Dad was down there one day on his horse. He had been in the orchard and rode on down to see about the crew planting the right of way. He tied his horse in the shade across the road. Some old lady walking to Ronceverte stopped, untied it, got on the horse, and rode off! An hour later some man stopped and said, "Mr. Cary, your horse is going down Ronceverte Hill!" Dad knew about who the old gal was, so he sent one of the men back to the barn for the car and down the hill after his horse, found the horse tied at Mrs. Vaughn's gate. He put Henry Austin on the horse. He rode the horse to his house, kept him all night, and rode him to work the next morning. Pop took a lot of ribbing over that horse.

It was a horse and mule powered farm up until 1940, when Mr. Ed Knight modernized the farm by purchasing the first John Deere tractor and a Caterpillar crawler. The forties rolled around faster than you ever knew. By this time, after the death of Mrs. Knight, then Granny Paxton, then Mr. E. W., things had changed a lot. Ethel was making her name in Charleston. After her father died, she never came back to Tuscawilla to stay. I only knew her to drive through twice. The orchard was at its best, the hog herd was big all over the country, and Shorthorn cattle in a big number had taken over. Pure bred sheep were here, the rest still going. The gardener had retired, so a few new faces were around here. We were into showing cattle and hogs, and that was what we kept with mainly through the years Ed, and then Julian ran the farm. We still have hogs in the pen just outside my bedroom window, but nothing like what I remember as a kid.

Carys of Tuscawilla

This would have to take you back eighty or more years. The first caretaker I can remember hearing about was Eli Blake, then Nip Cartmill. My father, Frank J. Cary, worked for Mr. Cartmill on the farm as a young man. His job was caring for horses and driving Col. and Mrs. Eugene Dana.

Mr. Cartmill was here seven years thereabouts. Mr. and Mrs. Cartmill had five girls at the time. The farm only had two houses, the brick house and a three-room log cabin on the back side in the hollow. Back in those years the farm manager lived in the brick house nine months out of the year, and when the Knights came for the summer the manager's family would move to the log house in the hollow. Over the past years I have been told by many folks about all the parties and square dances the Cartmills had in the wintertime in the brick house. They always said Nip Cartmill was looking for husbands for his daughters; not a one of them married in Greenbrier County.

Our father did some of the driving to the Greenbrier, Sweet Springs, to Salt Sulphur Springs and other places. When Mr. Cartmill left because he wanted better things for his girls, they moved to Point Pleasant. They were there a short time, a couple of years; from there they moved to Missouri, where he managed a bigger farm. From there they moved to Burford, South Carolina, where they ran a hunting lodge. She fed the hunters, and he was the guide. They hunted mostly deer and waterfowl. Both lived to be in their eighties. How I know this is our families were good friends; my dad went to their lodge to hunt three or four times with a group of men.

When the Carys moved up to the manager's job, it was the same as for the Cartmills; they stayed in the brick house nine months out of the year, then stayed in the three-room log house on the back side of the farm. Mother and Father were married March 28, 1915. Dad's mother, Grandmother Cary, lived with them; she stayed in the log cabin the year around. She was a widow; Grandfather had died in 1909.

In the late fall E. W. Knight would do a lot of hunting here on the farm. He would bring friends with him, he was a great bird hunter, loved his bird dogs. Mother would do the cooking for the hunting party. The Carys were like the Cartmills: they had square dance parties and such, and skating on the pond.

The first Cary was born in the brick house was my sister Margaret, in 1917, then a boy, Joseph Franklin. He lived only a short time. By then our house, the house we have lived in for so many years was planned and started. Ed, then me and then Julian were all born in this house. All four of us kids lived our entire lives here on this farm, working for and with the Knights, and many, many other fine folks.

King Apple

There have been many, many, many apple trees on this farm over the years. They grew five thousand in all. Apple was king here for many years. The first orchard was planted on the back of a very steep ridge looking east. In all, the orchard covered 50 acres and only a building stood in the way of a complete row of trees. The old orchard was planted in around 1890; it was a small family orchard. The new orchard was planted in 1923, and for a great number of years thereafter all missing trees were replaced. The back, looking east, was a very steep ridge. The front side looking northwest towards the main route to Route 219 was rolling to flat land.

The first thing the new orchard changed on the farm was that a new water system had to be found. No longer could the three-farm cistern and one well furnish the water that was needed for the orchard. First thing was to locate the best place for a deep well. I was always told that a Mr. Legg used a peach branch to find the water. The first well was on the back side of the orchard at the foot of the steepest hill. Ott and Eagle put the first deep well piston pump with a gasoline motor in somewhere along the line. I remember hearing that he had to drill thirty-six feet for a good flow of water. Behind the first pump house was a good sized overflow pond. It was a great place to play. Some of the men put goldfish in the pond to help keep it clean, lots of goldfish over the years, and the biggest bullfrogs you ever saw or heard. I always rode the old gray mare to the pump.

After the pump was put in, then a large wooden tank was built on the highest point on the stretch overlooking the orchard, the rest of the farm and surrounding lands. Water lines were laid from the pump to the tanks, from tank to barns and houses and the spray house. Edgar Ford built the first tank, and I think Mr. E. Blake was the first pump man; he lived right nearby. Speed Arbaugh was later, then Joe Cary, later on Charley Hughes, and then there was my dad, Frank Cary and my brothers Ed and Julian. I remember several fine fellows who, over the many years, repaired the bully thing: Mr. Plunkett from Lewisburg, Mr. Cable from Ronceverte, Mr. Arbogast from Ronceverte, Mr. Huffman from Lewisburg, Mr. Patton from Lewisburg, and Bill Ramsey from Fairlea.

I can remember more than once of being in the way, setting up half of the night at a fire behind the pump house while the man pulled the rod to clean the mud out of the well. Once or twice, we had the well digger back to dig a little deeper. In 1945 a new well was dug eighty yards below the old well. A new pump house was built as well, a new electric pump put in and a new water tank. The tank came by train and a P.O.W. helped the day and a half it took to install it. For many years, every spring it was drained and cleaned. A ladder was put inside, and it was hand scrubbed, rinsed and four or five gallons of bleach was dumped in. This was Lynn Green's job. When the new barn was finished in 1921 there was a big new apple room called the grader room. This was a long narrow room on the top side with lots of windows on the back side and one window on the side. It had three double sliding doors at the front facing the orchard right beside the hay mow and had a pair of steps on the ground underneath.

The orchard was the biggest part of the farm for many years. Planting, pruning, spraying, replanting them once in a while every few years, thinning and the picking. The orchard came into its own around 1932. My dad was a pretty fair orchard man. His brother Jim, and John A. Hamrick, Dave Deaver, Pete Lynch, Louis Winall, and B. E. Legg did most of the pruning for years. Lynn Green, and Mutt Morris later were the pruners. Jim Early, Harry Curry, Joe Arbaugh, Charley Sullivan were the cleanup crew. They hauled to sink holes and cut the biggest logs for stove wood. Joe Arbaugh was the teamster for this crew, and J. A. Hamrick was the true orchardman of this crew. He did all the grafting for the orchard; he was a fine orchard man. He was from New York state, part German, and made his home in the old part of Fairlea. Anytime the crew could stand to work outdoors in the winter, they pruned the lane hedge. They also killed field mice by digging and with a dog, later with poison. They did a lot of rabbit hunting and this they did to keep the rabbit counts down. Mice and rabbits are great destroyers of young apple trees. Other winter work was repairing all apple boxes, ladders, wagon, and beds, repairing sprayers and graders.

In most cases, all supplies came in the winter for the next season. Most all came by rail to Ronceverte, first on the line train at the end of lane, a couple of cars of baskets, fancy paper, linens, caps, and cardboard boxes. Most everyone helped to unload: the railroad gave you only a short time. It wasn't an easy chore, but one most all liked. We had to store most of the supplies in the carriage room.

Before the day was over not all but most of the crew would put all their change together and someone would hike up a street or two to a bootlegger to get a couple of bottles of their cheapest wine. Later they'd go to the state store.

My dad did all the planting or replanting of trees in early spring. Most of the time the trees came real early in March and were stored in the root cellar and were planted any day the ground and weather permitted. He only liked bare rooted trees and used six- to eight-foot trees. The rows had to be in a line. They used a string for a guideline at all times.

Early spring was time for the first of many sprays. It was an oil of some kind to kill all the things that had lived through winter. The first sprayer I remember was a big wooden tank sprayer, pulled by horses. It used up many good horses. It took four men to spray and one to drive the team. Allen Yates was one driver, Joe Arbaugh was another. One man had to be on top of the sprayer most of the time. Usually, it was Uncle Jim. It was his job to spray the tops of the trees on both sides of the sprayer, two rows a trip. Two men behind on the ground each sprayed a row.

The weather had lots to do with the spraying progress. Wind was a big factor here, and you always had to keep an eye on the rainstorms. For many years all the spray was mixed and made here on the farm.

Out in the orchard was a two room house, called the spray house. The smallest room held all the spray supplies and all the spray clothing. The other room held a big five hundred-gallon iron tank with a fire box under it. Most all sprays were cooked for hours, then cooled and mixed with water. Uncle Jim Cary did the cooking when not spraying. Harry Curry also cooked the spray. Thousands of gallons were used spraying each year. The men who sprayed had skin that looked like saddle leather from the wind, the sun, and the spray.

The back side of the orchard was so steep sometimes it took four teams to pull the loaded sprayer up the hill and also to hold it back going down. Sometimes when it was really wet, they could only haul or pull half of a tank. I know once they turned it over going down the hill in the last draw. No one was hurt but the best mule, Arbury, was never the same. Her shoulder was hurt. The teams used for the sprayer got the best of care. After each day's work the horses or mules were washed in warm soda and salt water. Their shoulders were washed in strong salt water, then fed well, and they got clean beds every night.

In and around 1940 the orchard got a new sprayer and a new Crawler-Caterpillar tractor to do the pulling. Cousin Joe Cary was the main Cat driver. Speedy Arbaugh, Clayton Deaver and my brother, Ed, all drove also. Many times they would load up, go out, start to spray, and the wind would start to get up. If the luck held, they could get the load off by spraying only one side of the trees, spraying with the wind.

When the orchard was getting ready to bloom another chore had to be done. The chore of placing the beehives around in the orchard. For years, the farm had their own bees, around fourteen hives. They were kept behind our woodshed. They were in behind a tall boarded three-sided fence. It was open on the east side; this was done to cut down on winter kill of the bees. Fodder and corn shocks were placed around the outside for warmth. The bees were fed sugar water in winter.

Back to the blooming orchard! Bees were left out in the orchard for two weeks. L. E. Legg was the old bee man, then Lynn Green. As years moved on the spraying materials became much stronger. It soon showed on the bees. Their number became smaller and smaller.

My dad looked around and found a bee man. A good friend over the years, Tom Dotson, who managed and worked for Nettie Jackson Dairy and also had his own dairy, was just starting out in bees. Ed and Julian, Tom and his sons Charley and Tommy moved them to Tom's farm one night after dark. For a number of years Tom and sons would case them up at dark and our family would go in the apple truck and bring them to the farm. We had around 24 hives then. After three weeks, Tom and family would come down and close up the hives once more. This he did by stuffing rags into the hive opening and driving them back to Tom's place. No spraying went on for three weeks. He shared all the honey with this farm.

Back to when the bees were behind our woodshed, so many times on hot days Jim Early, our faithful old colored man, a friend for many years, many, many years, always set on the old wood box on our back porch during lunch time, would sing out, "Bee swarm! Bee swarm!" Everyone would grab their buckets, dish pans, wash pans and tubs with sticks and start beating as hard as we could. The noise would drive them back to their hive. We would try to break the swarm and start them back to an empty hive. Lynn Green was the head bee swarm man by then.

By the middle of July there was sometimes another chore to be done. If your crops were showing too many apples, and you knew by picking season your trees couldn't carry these loads, you then, for your trees' sake, must get out your thinning clippers and go to thinning. You would leave the best apples, cutting off small apples, bad apples, and clusters of apples.

By this time of the year the farm crew had all they could do doing the farm work: haying, planting corn, hoeing corn, cutting wheat. We would go to street corners of Lewisburg to get a thinning crew of high school boys. They had to be young and good climbers. The going price was ten cents an hour.

Once or twice, we tried for larger crops so the farm crew went to the woods and cut props, pole props, and would prop up the limbs to keep them from breaking. Having to thin didn't happen so often. Dad always tried to thin a few favored trees for the biggest and best apples for the farm use. By this time of the year the chore of feeding the trees was always behind us. This chore was done by hand, using nitrogen of soda:

a farm wagon was loaded with bulk nitrogen of soda. One man drove the team, another man rode in the wagon and weighed the right amount for each tree and handed down the buckets. Five or six men on foot scattered nitrogen of soda. You started at the tips of the limbs going around the tree until you were at the trunk. I always hurried home from school on those days so I could drive a team while someone or two others would go feed and milk.

The old orchard and the little orchard were over by the Sullivan farm, later known as Fairlea, and the upper half was the very old orchard. Some fine old apple trees were found there. As the old trees died, they were replanted, and replaced by other kinds. Somewhere in my mind I recall there being twenty-five kinds of apples. The old apple orchard had some of my very favorite apples. There were four big Pippins – oh, what a good apple – a very large green-yellowish one. My mother favored apple butter apples. Ben Davis was a red apple; no one on this farm liked this apple, very white meat and very mealy. We had a couple of Summer Rambos, a late summer one, good big apple, liked this one to eat. We replanted with Summer Rambos once or twice and still have three trees now. One winter we had seven or eight big yellow summer sweets. I think it was called Pound Sweet. Everyone had a hankering for this sweet apple. It came in county fair week. We never went to the fair that we didn't take a basket of these sweets along.

My Dad's fairing started weeks before fair week when the first horses came in to train and work out. He also took apples to the stable boys every evening after supper. When fair week came, he always had the best parking place. He gave all the folks working the gates eating apples, all the parkers. He always got to go where he wanted to, all for a sweet apple. An old apple that I loved to eat was the Winter Banana, a very creamy yellow meat, one of the best of all keepers, this apple could be left on the trees up until the last day before the killing frost. It was a true keeper, would keep in the root cellar until spring. Another old-time apple, also a fine keeper, was the Russet Coats, a yellow mellow meat apple with a rough rusty coat. Later we had a couple of Rhode Island Russets. Same apple, new name. We had two Northern Spy trees. Pretty red, a good apple was the Roman Beauty. There were several Wolf River trees, another red apple, the largest apple of all apples. These trees were thinned every year. We always tried for the biggest apples, showed them at the fair. One apple filled a paper plate. This was a fine baking one and a great dumpling one.

All that could be raised were sold to the Greenbrier Hotel for many years. I can recall when they were showed or shown at the fair; folks couldn't get over the size of the apple. Loads of folks always wanted to buy them. I got to sell them on the last day of the fair. Could make good apple butter out of them too.

There were a few trees of Baldwins, only a fair apple, not much flavor. We had several Stayman, Winesaps – a great all-purpose apple, cooker, and eater – another red one. This was a good cider apple, one of the better apples for sale.

A good keeper, and the apple I would have to give the best all-around apple was the Maiden Blush, a very fine white meat apple with a yellowish ruby skin, a great pie apple, a good cider apple, and a first-class apple butter apple. It was a poor keeper; they came off weeks before the fair. We had to move them fast to market. This apple was an early bloomer and didn't hit every year.

We did have a couple of Sheepnose trees, a deep red October apple, and a Smokehouse or two. In the lower half of this orchard, you would find more Wolf River, Stayman Winesap, Russets, Ben Davis, Maiden Blush, Red Delicious, Golden Delicious and York Imperials or Johnson's Fine Winter.

The big orchard was made up mainly of York Imperials, a red winter apple and a best seller; it was the last one to be picked. It was stored for the winter and improved with age. The others were Delicious Reds and Golden Delicious, the Red ones were eaten, the Golden ones were good eaten or cooked. There were a few very fine Red Jonathan apple trees. This was my kind of apple, red and juicy, but not too large. But oh, so much juice! It cracked when you would bite into it and the juice would run down your chin. There were some Maiden Blush and a few Greenings and very few Fall Rambo apple trees. The Fall Rambo was a fine apple; it cooks well, keeps well and is good to eat. I am sure I could take you blind folded to where the Red Jonathan trees and the Fall Rambo were, and I am also sure the old bay mare, Princess, could take you there if she were still around as we made the trip almost every day when the apples were ripe.

The apple season started in August when a few bushels of sweet apples as well as the Summer Rambo were ready. Things got rolling into high gear after the fair started when the Maiden Blush apples were ripe. We sold most of them to fair goers and to the candy apple makers. The apple picking crew was made up of mainly extra help, with only a few in September. Many were here in October and November. The older farm hands were put in charge of these crews. Other farm hands had to harvest the corn crop and in general get things ready for the winter. Most of the trees were large and the apples had to be picked from ladders, wooden ladders. You needed an eight- foot ladder at least. We usually had fourteen-foot ladders. It was hard work: you picked into a canvas picking sack and once the sacks were full you had to put the apples into wooden boxes which were hauled on hay wagons by the horses. Each wagon held 44-bushel boxes of apples. One had to keep after the pickers, reminding them to pick easy, handle the apple lightly. We had young and old crews that picked up apples behind the others, picking up all the dropped apples. These apples were stored separate and sold much cheaper.

The picked apples were unloaded in the grader room where they were cleaned, sized, graded, and packed ready for sale. Here we had another part-time crew; this crew was made up mostly of women, usually a family job. There needed to be a couple of three to four working men to help. The apples were dumped by hand onto a roller ladder up on the grader, then cleaned and polished. The cider ones were all small apples and fell through the chain running to a table where they were placed into boxes, stacked, and then later were sent under the room into a cider apple bin. The grading was all done by hand, usually three to four women. They would sit on high stools picking out all bad, blemished, unsound apples. They had to be fast thinkers. The bad apples were sent down the cull side and then went up a size chain to a table. At this table you needed headers and fillers. Next came another size chain table with an opened side. The rest of the apples went onto the end of the grader where these number-one apples were packed into boxes. The cull apples rolled onto another table. This is where you wanted to only see a few apples, as the number of culled apples told you what kind of spray job had been done and how well your pickers did in not leaving finger marks on the apples.

The graders needed both hands to do the grading and sorting and they had to be fast. The work was hard, and you couldn't speak, rest or do anything as long as the grader was going, because it would swamp you and overrun you if you did.

The big number-one table needed three good, fast, headers which some folks called ringers, a tub filler, couple turners and a stacker. Most of all you needed sixteen good hands. Year after year the same folks worked. The headers or ringers were the ones who picked out the best apples, the reddest ones. These were put in a pan with the best side down, then a metal tub with a paper liner. When the tub was full it was slipped off very slowly and a new bushel basket was put on, then turned upside down, pan taken off and you had a super looking bushel of apples. This was the way you fooled the buyer with the best on top. We also packed a lot of fancy apples in waxed paper in bushel boxes, each apple in a section with colored fancy waxed paper. Most of these boxes were sold to the Greenbrier Hotel. Many of the number one apples were trucked to coal company stores in several cities. Lots of number one apples and fancy ones were put in cold storage in Ronceverte at RIPCO. All second and picked up ones were sold at the apple room and cider apples were used later for cider making.

Back in the 1930s, we sold many tons of apples to the United States government for the W.P.A. Program. These were stored in all of the underneath sheds. A man from the program was on duty and was sent out to four counties, when and where they were needed. Sometimes we sold apples to the Homestead at Hot Springs, Virginia, and to the C. & O. Railroad. In 1944 we packed apples for foreign relief. All the apples from this farm were sent to Belgium. We wrote notes, letters, and included pictures in with the barrels of apples. They were trucked to Ronceverte to be put on the train to New York where they were shipped overseas. We received a couple of notes back from the folks who received the apples and our letters. Charlie Hughes and my brother Julian did all the heading of the barrels. I can recall getting two loads, that is two boxcar loads, on the train. There were three bushels in each barrel. Joe Cary, Clayton Deaver, Speed Arbaugh, and Bud Williams were the truck drivers taking apples to the coal company stores.

When the apple crop was real large, we would put thirty percent of the crop in cold storage at RIPCO. We would fill two big rooms. The storage crew loved to haul to RIPCO in that other fruit was stored there too: grapes, pears, oranges. Mr. Guy B. Montgomery would always beg and plead for the guys not to touch the other fruits as it was for sale. But every time we stored apples there the crew would get into them. Mr. Montgomery got so he wouldn't let us in without one of his men along, so no one would get away with any fruit that didn't belong to them. Oh, how our crew loved the grapes at RIPCO! They always had three or four empty baskets with liners and lids on them. The lids were removed, and the grapes were placed in. Mr. Montgomery's crews had good times eating our apples. Over the many years the hauling crew, the loaders and unloaders drank tons of cheap wine. They wouldn't have two cents between them, but they would somehow get wine to drink. Many of the crew were folks from near here, they included the Brackman, Lynch, Winall, Morris, and Blake families.

The grader crew – oh there were many, many, many fine folks over the years – Toad Scott, Pole Hayes, Nina Green, Mary Green, Naude Dolan, Mary Frances Dolan, Sally Austin, Rachel Austin, Katherine Yates, Maxie Childers, Hattie Sanders, Eva Underwood, Lou Hughes, Marie Anderson, Ruth Anderson, Ethel Honaker, Christian Morgan, Nellie Anderson, Laura Burns, Betty Anderson, Lena Anderson, Mary Hammock, Helen Evick, Mary Folden. So many, it would take pages to name them all! How well I remember these fine folks. We had singalongs at noon and if anyone had a birthday someone would make a cake. Some of these folks walked three or more miles to work here; one or two walked five to six miles. Some were neighbors to each other; some were the best cooks. We shared with each other in happy times and in sad times. Some of our best friends came out of the orchard crews over the years. If there was a church meeting in Teaberry, the preachers would come up and have a little meeting, pass the hat, and get some apples. If we had filled a big order and the grading and packing had gone well and the apples were on the train, we would have a Coke and ice cream treat.

In 1944 we packed fifty-three thousand bushels of apples, which was state record for 50 acres of land. We also had a big crop in 1945.

In both of those years there was little male help due to the Second World War. The packing house was all women and we had lots of ladies picking. As November came along, we worked seven days a week and ran two grading room crews. On Saturdays, the state road crew picked apples and sometimes they worked on Sundays. On Sundays we let the whole family pick just as long as a man was with them to pick the top of the trees. For the most part the weekend crews were paid with apples. Sometimes they got ten cents a bushel and then five cents for a bushel of picked up apples; the poor devils didn't make much. Both years we had P.O.W.s from Ashford General Hospital Prison camp over at the Greenbrier Airport. We worked 30 to 70 men; it all depended on how many we could have and how many other farmers needed them. We found most every man was from the same section of Germany, a farming section. We had only what we would call two bad Nazis; the rest were quite happy, glad to be here away from the hell in Germany. After the war, we continued to have less and less apples, but still were able to make and sell cider until the 1960s.

In the late 1940s and 1950s the cost of labor and supplies were too great; the orchards were let go. All the trees were cut for firewood, never sold any, just for the help and friendship. The stumps were pulled, the land was turned all, the front to crop land and the back to pasture.

Fruit Inspectors and P.O.W.s

As a major producer of fruit, we had many U.S. government fruit inspectors on the farm. Part of the day they inspected here, then over at Sinks Groves. Some years when they were short of inspectors, we only got spot inspections, once in a while. Like in all walks of life, some were fine fellows, some were just plain kooks!

I can recall a few: Bill Worthering, a young fellow, he stayed here with us. We had to feed him; he had an eye for the ladies. Martin Rodgers from Organ Cave, a local, his eyes were made for something other than ladies: booze! Then we had a redhead, Sandy Perkins. He was from Maryland and did all his work from the pool hall in Lewisburg. B. E. Eppling was the kook of all kooks. I all but cut his ears off. Man, he was a troublemaker from the word go! He ran every young girl in the grader room. Took a different one out at noon time every day and he wanted to take one to Neff's Orchard every time he had to go.

One day he had plans to cat around, but I had plans first. It was homecoming at the high school, so I left five minutes before kick-off and was back five minutes after the last play. When I got back to the grader room, Uncle Jim was in a mad rage. That son-of-a-gun Eppling had thrown out two hundred bushels of apples. He'd said they wouldn't pass inspection. I got hot in the collar and asked what was the reason for them not passing?

When we stopped working at five, I asked four of the crew to stay over, and with the help of the clean-up crew repacked those same apples. The inspector tore the baskets full of apples up so much we had to re-pack. We took the lids off of the baskets ran the apples back over the grader didn't find a bad apple or wrong size apple. We refilled the baskets, turned them, capped them, and restacked. This was Friday night. We worked three hours, sent for supper at Oscar's, and worked a couple more hours. When Bill Eppling got back at noon on Monday, he took samples and inspected the unpassed bushels. The bushels ran that morning. He didn't throw out one apple, not one. By Friday word got out to him what we did. He didn't show up anymore. Someone put the skids under him, wonder who?

The head inspector for many years was Woody Reynolds from Inwood. He came by once in a while and had to work ten days on and off when they were short of inspectors. He was a fine man, spent long hours teaching Julian and me all about the orchard and grading room. He stopped in for years after we stopped having apples. He helped to get the state marking off the ground. They had a potato grader here in Fairlea and at the fairgrounds. We helped him to grade potatoes once or twice.

I mentioned earlier that we had P.O.W.s working here. As I recall, we were real lucky. We got along fine, had no troubles from anyone. We found most every man was from the same section of Germany, a farming section.

The way the prison workers worked was this: you had to haul them to work, but with the orchard truck hauling apples every day, we had to use pickup trucks. Julian and I did the hauling. He drove over, picked me up and dropped me off at Merry Hill Farm across the road. I drove their truck. You had to be there in line by seven-fifteen in the morning. I saw the same trucks there every morning: Gene Knight for Valley View, Uncle Joe Hayes for Greenbrier Farms, one from Nelson farm, and some hauling services for three up around Williamsburg, Nell's Orchard and sometime RIPCO.

Tuscawilla got P.O.W.s five days a week, the rest one or two days. Gray Rock had them every day also. Most days we got only seven to twelve men, but we got the most. We had a bad feeling about the P.O.W.s at first, with brother Ed over there and us not knowing how he was or where he was. We would get one guard: he carried a sub-machine gun and was supposed to ride in the last truck, but seldom did because I was always last, and the truck I drove didn't have a heater! Julian had the first truck and would haul the biggest bunch, but sometimes when we had a big crew, the loads would have to be the same. I broke down one morning, one cold, frosty morning at the foot of Cabin Hill. They all got out, talking, using their hands to see if they could fix it, but we couldn't get it going so they pushed it up and down to Pine's Curve. We all got in and we met Julian in Caldwell. He had been to the farm and unloaded. The guard was ash white, and he rode in the last truck thereafter.

The P.O.W.s knew all too well how well off they were. The guards would always go to hay mow or the garage apartment and sleep, and when it was time for their hot lunch to come, a P.O.W. would go find the GI. Once in a while they would pull an inspection, and the P.O.W.s would see the Jeep coming. They would sign out and the one P.O.W. could go get the guard up and on duty.

We always let them take back to camp all the apples they could carry. This went on for a spell. The working P.O.W.s had to take them in to the confine and turn them over to the officers who did not go out on work detail. The officers got to stay in the mess hall after chow to play cards and when the others were locked up, the officers made booze out of the apples with spices, yeast, sugar, etc. Then one night there was a drunken brawl in the officer mess. No more apples. They were searched at the gate. Only one apple in each hand from then on.

We went over one morning early, real cold, real frosty, and got in line. I was the first. We sat there a while and could see things weren't right. A guard, a sergeant I knew real well, came by. He said things were bad and he didn't think they would let them work. Uncle Joe and Julian pushed my truck back down the line away from the gate. Everyone was armed to the hilt. It was all Army.

We soon found out the guards had just killed two men on the coal pile: the mean P.O.W.s were kept on the coal pile. One P.O.W. had hit a guard with a shovel and the GI shot him. The others got in the fight, and one was killed. The first one died an hour later. We were told to go home real slow; trucks were searched at the hard road. They didn't let them out for four days after that.

One P.O.W. was a bad egg and we had to keep an eye on him at all times. He started a fight in the orchard with a fellow P.O.W. He would come up to the packing room and talk to the women. He got a pocketknife one day. They found it on him in the search that day. He went to the coal pile.

We had a good artist. His name was Bill Hayleafe. He painted us all a picture; he was a farmer with a crushed hand. We got his paints for him, and he used boxes to paint on. He had a wife and two daughters. He wanted to come back after the war, a couple of others made official requests also. One of them, a man called Alfred, corresponded with the folks he met here for a good many years after he went back to Germany.

This was all years ago of course, and just recently Kaaren's husband Larry's sister-in-law – his brother married a German lady –

brought over her nephew, who's interested in farming and was over here visiting. Kaaren and Larry brought him up, and that boy didn't speak a word of English and I don't have a lick of German, but with poor Sabine translating like crazy I showed him around the place. He was interested in the tractors. And I never would have thought that someone like me who's never really left this farm would have met so many German farmers!

Cider Making

Big frost this morning! Now it's time to crank up the old cider mill! We never liked to make cider until after the first frost; that's when the apples had such a rich color. October was our month for cider. September was too warm, too much other work to get done. We had some great times making cider.

At first, we only made it when someone or some club or hotel wanted cider for a party, just small batches to drink. We used the old hand-cranked screw-top muscle powered kind of cider press. It had two tubs: you filled one tub with chopped apples and moved it up under the press. The first dipper of cider was good sipping. We tried hard to always have some eating apples mixed with cooking apples for most ciders, but for a tangy cider we used cider apples, which are the smallest apples.

When we started the mill in the fall, school kids would come out and make a batch for a party or ballgame. When my classmates would get together someone would always bring up cider making at Tuscawilla and my dad. What good times we had! When we got to making it to sell, we got another mill just like the first one. It took lots of muscle power to run off a big batch like five hundred gallons. Old Jim Early could crank off more cider in a day than anyone I ever saw.

We washed the apples and drained them. We caught the cider in buckets and then ran it through cheesecloth into another bucket, then funneled it through another cheesecloth, straining it into a wooden sixty-five-gallon barrel. These barrels were scrubbed and let stand in the water to swell, then the wood was re-waxed. The cider barrels were kept in the root cellar to keep cool.

The cider for farm use was a sweet cider made into vinegar, and there was always a barrel of hard cider. If the cider was for sale, we would bottle it in gallon jugs. We used second-hand jugs which were washed and used the next morning when the fresh cider was jugged. A label was put on each jug. A few years later we got another mill, a big one. This mill was run by a big electric motor. It had a conveyor belt which took the apples from the drain table up to the top and dropped them into the chopper box. They were chopped into burlap sacks, and these were stacked seven high when they were pulled under the press. The apples had been washed in a big cattle trough and bucketed onto the drain table. As the cider came from under the press it ran into a big bucket and then pumped into the barrels in the root cellar by a rubber hose. We could make five hundred gallons in three hours.

The jugs were washed in the dairy room, and the cider was jugged the next morning as early as possible and then packed on a truck and sent to Charleston to be sold in several stores. The last week of October was the busiest time. We had to make cider twice that week. There were some cold times around the mill; sometimes so cold ice would form all over the mill. One time in October there was a really bad ice storm and we had to have a big load of cider ready to go to Charleston. Only those of us who lived here on the farm made it out to make cider since no one else could get here. It was a bad storm. Julian and I went out and looked over everything, seeing nothing hurt and the livestock okay, we did a fast-jugging job. By nine in the morning, a couple more workers got here. Julian and Ben Hayes started to Charleston with a big load of cider. It took them seven hours to get there.

We made lots of cider for the Greenbrier Hotel, lots for the high schools and some every week for Greenbrier Military School. Miss Willie and Miss Kate Moore loved to have cider. In the years when the apple crops were good, the high schools got to make cider for free. They just had to have lots of help and to clean up.

We didn't always use cider for sweet cider; every year a barrel was made into hard cider. Sometimes it was put into a seventy-gallon glass jug. Man alive, this would set a few on their ears, it was hard stuff! The men cooked potatoes, ground barley, sugar, raisins, yeast, and cider for the hard stuff.

Once or twice all who partook in this hard cider got so high their family had to walk up to the farm and guide them home. One fellow, Garnett Anderson, got lost in the Sullivans' corn field. He lived in Fairlea and his path to work came through this very field. This day, on the way home he got out of the path and went down in a row of corn. He had just been married, and his wife and mother came on the hunt for him. They found him by sound; he was groaning, and everything. They were all half dead the next day after having hard cider.

We would also make a barrel of vinegar for the cooks. The hard cider and vinegar were made after the last making of the year. Over the years we had a mountain of mashed apples. We fed this to the hogs, we put it into the manure spreader and spread for fertilizer, used them for mulch around shrubs, we gave them away, even sold a load or two.

Apple Butter

When are we making apple butter? Are the apples ready? What kind are we using this year?

October was always apple butter time here on the farm; just before the leaves came down, as I recall. It was harvest moon time: pretty sunny days, cool nights, and early frosty mornings.

Being a good-sized apple orchard, apples were always plentiful, and we could make apple butter from any season's apples: summer apples, fall apples, or winter apples. We never knew from one year to another what apple we would use. The old folks all had their favorites. We liked Rhode Island, Grunings, Pippins, Fall Rambos, Northern Spy, and big old Wolf Rivers.

We tried to make apple butter on a Wednesday; if we were rained out Wednesday, it would give us a couple of more days. We had an old-time apple peeling the night before. Lots of folks would come, all with their favorite knives, some also with their favorite pie or cake for later. Coffee pots were always on. The Joe Hayeses: Uncle Joe, Aunt Mary Rose, Eva, Pole, Susan, and Helen; the Jim Carys: Uncle Jim and Aunt

Fannie; Aunt Bert Jones; the Will Hayeses; Uncle Bill and Aunt Mary; Connie Conyers, Aunt Lizzie, Nellie Anderson, and many, many old friends over the years. Uncle Joe and Uncle Jim would do all the peeling, and the rest would core and slice.

Never could tell how many apples it would take. It depended on how good the apples were: how dry they were or how juicy. I recall we would always have six full bushels of cut apples and be ready to cut more if needed. The apples were put in clean bushel baskets which were covered airtight. All the time they were peeling, coring, and cutting there was much talk of older days, older folks, and many other apple butter days. When all was ready for the next day the kitchen cleaned up, jars carried in, and a big cake or cobbler was brought out. The best kind of eating from some of the best cooks around! Then we all went home to bed for a night of rest.

We always tried to have the apple butter cooking by seven in the morning. If it looked like rain, a make-shift cover was put up over the kettle. The copper kettle was always cleaned that morning with vinegar and soda. It would shine so you could see yourself! We used a thirty gallon kettle.

We had lots of folks to help stir: Jim Early, Lynn Green and Austin Jackson. The apple butter firewood was laid, and we always put about a gallon of cider or water in the kettle, then the fire was started. As it heated up the stirrer was put into the kettle, then a bushel of apples was put in. They had been washed and some water clung to them. When the first apples were in the kettle, the stirring started and didn't stop until the kettle was taken off the fire many hours later. We always put copper pennies in the apple butter to keep it from sticking. You had to stay with your fire all the time – not too little, not too much – just the right amount of flame at the right place under the kettle. One man was the head stirrer, and the ladies would help. We sang a little ditty while we stirred: once through the middle, once around the outside, once back around, once more through the middle. If all the apples were in the kettle by three in the afternoon or thereabouts, you could hope to get the butter off by five-thirty.

We always had big eats for all: backbones and potatoes, chicken and dumplings, green beans, late corn on the cob, peas, all sorts of other things. In the last hour we had to put the sugar and the seasoning in. This would make it get a much redder color and start popping and bubbling, getting thicker and harder to stir. By this time, our eyes would be full of smoke, our hair full of ashes and our taste buds overworked!

"Oh, don't know, a little longer, just a little more! Is it ready? Not yet, just a little more will do it." That is what the menfolk wanted to hear. The jars were washed in the kitchen on the stove. They were kept warm and carried out in boxes so the cool air couldn't get on them. The fire was dug from under the kettle and put out so the smoke would go away. A table was moved up nearby the kettle.

And the dipping started! Jars were filled. Right tasty! Lids were put on, jars cleaned and put back in the boxes. All the time folks were looking for the coins. Sometimes they got into the jars and had to be spooned out. Just like new, the coins were so shimmering! There wasn't much supper as there was too much tasting, too much lunch!

Around the apple butter kettle year after year, there had been a lot of stirring, fetching, popping, bubbling, smoke in our eyes, ashes in our hair, and loads and loads of happy fellowships. What a happy time apple butter time was! I recall making it for years, sometimes here at home, sometimes at Uncle Joe's, other times at Uncle Jim's. It never changed: same apple peeling, same stirring, old faces gone, different faces there, younger faces, someone always gone from year to year and someone new to welcome.

Uncle Joe wanted to use only one kind of wood, poplar. He would try and cut one about a month before we made apple butter. Up at Uncle Joe Hayes' one day we used Wolf River apples. By four-thirty we had no color; still pale and white. I went to town to the five-and-dime store and got six bags of cinnamon red hots. When I got back to the kettle, and when no one was looking, I dumped in a couple of bags and had to stir hard! It wasn't too long until our apple butter was glowing with a suntan, a pretty red apple butter with a nice cinnamon taste.

In later years we moved from Uncle Joe's to Uncle Jim Cary and Aunt Fannie's. Lots of folks were gone but we still had the guiding hands of the old folks, and the older they were, the harder it was for them to agree on the making. Apple butter days were good days. It was very hard work but so rewarding. We loved to get together, loved to hear stories of older days, family history, and their happy times. If we had only listened and learned. They made the very best apple butter.

We tried back here on the farm again, after all the older folks were gone. Lynn Green, Sally Austin, Ruth Smith, Mary Edgar Hayes, all did a pretty fair job; it wasn't like the old folks, but it was good. We tried here on the farm once to make it out of cider; were going to put it on the market. We used pint jars and made it here behind our wood house with the cider making gals helping. We had one thirty-gallon copper kettle and two fifteen-gallon ones. The cider was made the day before. We cooked all day; the longer we cooked the bigger the mess got. We just couldn't do a thing with it; it got so thick, so tarry. After a very long day not knowing what we were doing, all with different opinions, we finally put it into jars. It was very tasty but looked inedible. We labeled the jars, cleaned them, boxed them, put them on the cider truck the next morning for the city market in Charleston. They didn't send any back, but they didn't order any more, that's for sure! We found out later we needed steam heat; it would have been much better. But we tried, and I am sure all the old folks would have frowned on us for the waste.

Horses

Horses were king for many years, many, many years. Countless horses found their happy homes here on the farm. My dad came to work at Tuscawilla as a helper in the stable. He had been working at the Stratton-Sydenstricker livery in Lewisburg. He did a lot of driving for them to White Sulphur Springs; drove rigs, sleds, buggies, and the last year there he drove a coach to Sweet Springs. So, he was a young man with lots of horse sense when he came to Tuscawilla. The place was loaded with good horses: they were used daily for labor and pleasure and fun. There was an outstanding horse farm across the road from Tuscawilla: General Watts had a big horse farm with a barn and a half mile track.

The first horses I recall hearing the family talk about was an excellent team of bay driving mares, all class, top-notch. They always talked about driving to the Greenbrier to dances and parties, and how this team of bays always led the way home. Their names were Baby Gale and Sissy Bell. They were not sisters, and both were bred to Watts' trotting horse called Watts' Red Baby. Out of this came a big driving horse called Halstead, a big reddish bay. He was the main buggy horse for years. The other man, Mr. Watts, had a small mare colt called Watts' Princess.

I know by stories they told that the Watts horses would race with any horse who stopped in on a weekend when the Wattses were home. They were raced during the week by the help: Mother's brother, Will Hayes, lived on their farm and worked the horses the same way as my Dad on his side of the road. I can remember the working horses, Maude and Hade. I don't remember their colts. Both were full sisters, both matched speckled greys. The last colts they had were mule colts: one a guy mule called Tazar, the other a big bay mule colt called Arbutus. Both lived their full lives here on Tuscawilla. They were the pride and joy of Dad and Mrs. E. W. Knight. The others were Old Ned, a pure white horse; not grey but pure white. Dad rode him in a bunch of Shanghai parades. He was a quiet and gentle horse. Nellie was a big sorrel mare; he got her from a horse trader from North Carolina. She got a bee sting one day and ran off with Uncle Jim in a single horse hay rake over in the little orchard. They tore up the rake good and Uncle Jim caught hell from his baby brother!

I knew Old Ned; he was old but was raised on the farm. He was an old fellow who knew what to do by himself. He was my ride when we pulled hay shocks. Most of the time he was Bub Legg and Lynn Green's plow horse for the vegetable garden. He was a pretty red horse, easy to keep. I saw him have the colic one day at noon. Dad made the men run him up and down the hill behind the barn. Dad had a whip, and he busted his tail and all but ran the men to their knees – must have run him two hours – he got over it and was okay. When he retired, he was taken to a home in Caldwell and was used only to plow the vegetable garden. He died there and Dad took a team there and dug a grave.

We had old Bob, he was a lazy one – Julian rode him to the hay shocks – you needed a good stick on Bob and had to use it. He was a dark brown. Gail was a very small sorrel mare, got her from a gypsy horse trader to plow the vegetable garden after Bill retired. She wasn't much. We had a black mare called Jane and a big bay mare called Myrtle. They both came from a North Carolina horse trader.

I recall Dad was cutting wheat with an Ott binder and an Ott team. He had Nellie and Ned as the second team. He hooked Jane in place of Ned and told the horse trader to come back in a couple of hours and if Jane was still afoot, he might trade. He did trade a riding horse called Jingles and paid boot for her. A couple years later he bred both mares to a reddish grey stud. Myrtle had a big bay colt, a mare. I got to name the colt, and I called her Tea. Charlie Hughes named the other colt Coffee, a black one. They made one heck of a good team. They started to Caldwell, to the stockyard one Saturday morning with Charlie Hughes riding one and leading the other. They had their harnesses on, good harnesses. They waited on brother Ed to come along, and a man, a Taylor from Renick, made a deal with Ed, so they sold them. He backed up to a knoll and loaded the team and away they went. He paid twenty-five dollars down, never got another nickel on them. Ed and Mr. Ed Knight went to see the man a dozen times. One day Mr. Ed just gave the man a paper he had put down, paid in full. He knew they had a good home, so he let it go.

I have about run out of work horses, but not yet. I left out the matched grey mules!

When Tuscawilla was going to log the woods, they asked Julian and the Fil Arbuckle cousins to see about a matched pair of mules to log with. They were taking a couple of horses themselves to the sale: a horse and mule sale no less. They came home with pretty matched dapple-grey mules; paid around eighteen hundred dollars for them. One was Hank and the other was Ruben. Someplace else I may have called Ruben something else, some other name, but it was Ruben. They turned out to be bad news; we couldn't do a thing with them. Charlie tried so hard; Ed helped, and Julian helped. The one called Hank, they couldn't get a bridle on, they had to use a ladder and needed three men. So, the Arbuckle cousins went back to town with the mules and harnesses; they only got twelve hundred dollars for them.

There's one more mule – don't go away by golly one more – ha! This was a very small grey mule with some age on her. We called her Tinny. We were putting up hay over at Smith one summer, down at the bottom of the hill by the cabin building. We needed three stacks of hay for the winter feed for the horses on the pasture over there. I rode the old grey mule that day – all the crew was there trying to beat the rain – and the old mule ran over a bees' nest in the ground. She was stung all over. I got hit by a dozen or more. Bud was coming to help milk and had to take me home. They put the old mule in the paddock. They got the hay stacked. Dad came home, got a bucket and baking soda, and went back and washed the mule for a couple of hours. They left her over there overnight. Mother washed me down with soda and put cold cream on me. Lordy, Lordy!

The last little mule, Tinny, was the kickingest mule of all times. Just start to walk behind and she would kick with both hind feet. Uncle Bill Hayes was looking at the mule one evening when she kicked several times. Julian said, "Yep, she'll do it every time, be careful."

Back to the riding horses. They rode for fun, rode in fairs, at horse shows, camping trips and hunting. Halsted was the first. He was shown at the Greenbrier Valley Fair in 1923 by Adeline Dana, Ed Knight's cousin. Her sister Arnold rode the same horse at horse shows at the grounds at White Sulphur Springs and also at Hot Springs. Prince was an old timey horse; rode until he was worn out; so sway back. Princess was driven and driven. She was a real easy horse, a pretty bay mare. She was the last buggy horse, and she was the one I swam the Greenbrier River on a couple three times, and the one who took me over Coffman Hill to Big Draft Road up to the very high top. Ed Knight and I did this three times. It was an all-day ride and there were saddle sores, oh! How there were. My brother Ed rode his horse, a nice strong one. Princess was the cook's buggy horse until she retired. Rachel Deaver and her dog Trixie went to Fairlea every Thursday afternoon to her little house and back before bedtime. They also went to the Presbyterian Church in Fairlea every Sunday.

Then there was the big horse, the reddest horse I ever saw. We called him Red Eagle. He came from the stable at the Greenbrier Hotel. He was Ed Knight Junior's first horse. We always used him for the cows. He was used to pulling a two-wheel cart. Every three weeks Mr. Ed Knight and Ed Junior would take dog food to Bethel where Lynn Green lived. Lynn and Mr. Knight had dogs together, hounds and bird dogs. I suppose they were Knight dogs and Lynn took care of them. They also hunted together. The hill behind the barn was so steep after a few years Red Eagle got a stiff neck and couldn't get his head halfway up. Julian hauled hay shocks on him. Julian was a little fellow and all you could see was a hat and a horse.

Jim Peyton came to ride with Ed Junior when they were 15 years old one July day: red hot and were told by Pop to ride slow and easy in the saddle. Ed Junior came back to the barn half an hour ahead of Jim. Dad made everyone look after their own horse, saddle, cool them off. Ed Junior had his horse ready for the stable when Jim came in on an almost down horse. Old Red Eagle was all but gone. Jim dropped the reins and wasn't about to put the horse away. Pop all but tore the barn yard up. That was the last time Jim ever rode a horse on Tuscawilla. No one ever mistreated a horse on this farm or any other place if Pop was around, no one.

There was a bay, Bobtail Bess, a buggy horse and rider. Old Bess and her mother, both were bays. We had old Halstead, a bay buggy horse, and young Halstead, a very high-spirited riding horse. He tossed Eddie – that's Ed Knight, Jr. – two or three times. He never got hurt, only his pride. Aunt Ebbie Knight had five horses they showed. Once or twice, she showed in Virginia, at Hot Springs. The first two were both jumpers, in fact all were jumpers. First was Orange Glow, a light sorrel, and Blue Moon, a very dark grey. Then we had Highball, Rye, and Barronwood. He was the first Aunt Ebbie rode, a fine looking dark horse, a dapple brown I would call him. Highball was the best-looking horse ever around here, a good jumper. Dad raised him. A young good-looking boy named Marshall Bond who I might've mentioned rode and showed the horse. He did well. Rye was a big, big grey horse with red. Aunt Ebbie rode him in hunts.

We had a black horse, Pete. Everyone loved him. Pete got a bad name: he pitched Mrs. Gertrude Parlier off down in the front field. Her twin sons were riding with her: one on Rye, one on Gaylord. She hadn't been on a horse for years and they raced, and she couldn't stay on. Her back was broken, and she was down for six months.

There was a grey mare, she was gotten from Bert Bobbitt for Mrs. Lucille Knight. She didn't ride but a couple summers. We had a little sorrel mare, a black one, a poor horse not sure footed. We got an old mare trotter named My Jane. She had been on the track but didn't set the world on fire. She was a small track racehorse and was overused and over doped; she'd lost eighty percent of her eyesight.

Dad had his own horse, saddle, and bridle. The horse was Barronwood's half-brother, his name was Sir Barne, a little bigger than Barronwood, you could ride or drive him.

The carriage house, which was really the big three-door room by the hay mowers, was where the sled, hack, buggy, carts, and gigs were kept. The saddles and bridles were kept in the horse barn, then later in a saddle room in the dairy barn, which was a first-class one.

We had lots of foals to cuddle on this farm. Men who drove the work teams were Allen Yates, Jim Cary, Charley Hughes, Henry Austin, Stud Brackman, Ed and Julian. I drove lots. I drove when we picked rocks off the vegetable garden every year. Fertilized the orchard, hauled apples out of the orchard, picked corn, etc. Henry Austin was bringing a load of apples in from the orchard, he was driving Old Jane and another black mare called June Bug – she was a new one – just traded for her – they were in the fourth apple row from the barn when the ground fell out from underneath them. They were in a hole over their heads, both on their feet but broke the wagon and some of the harnesses. Didn't hurt Old Jane but it tore June Bug up; she was turned out and traded later.

Over the years Pop traded horses with anyone. There was a bunch of gypsies that came twice a year. My favored one was Ma and Pa Tom Farley. They were from Salisbury, North Carolina. We would go see them every evening they were here. You never saw such rings they wore. They always stopped at the Lewis Springs, they may have stayed a day or two, then some ladies of Lewisburg would have them run off, but they would just go outside of the town along the road. They had a horse trailer they pulled. They would park on the road behind Greenbrier College; Pop would go by every evening and on Sunday. That was the first home on wheels I ever saw. And I'll say it again: they wore the biggest rings I ever saw. He would burn Pop once and a blue moon. They always had a couple of colored boys with them, had one called Jinks that could dance and also played a mean banjo. Professor Charley Echols, John McClung and Claude McLaughlin went by about every night.

Dad would make them bring the horse and if he or she worked well and wasn't winded he usually got one. We traded a bay mule to them that was raised here on the farm. Their barn burned that fall early, burned twelve head of animals. Old Arbutus, the mule, was one of them.

This farm sure looked after the horse manure, and there was a lot. First of February Lynn Green would start piling it outside the barn door, by the eighteenth he would have had all over behind the apartment. He had three hot beds along with a cold frame. He would redo the dirt and pack the horse manure around the hot beds and plant his seeds and horseradish bed, asparagus and rhubarb and mint beds which he worked the manure in and around the beds. He used everything. All the straw bedding from the horses was used to cover the strawberries and put around the raspberries.

Those were good days and good times. There's nothing like a horse!

Hired Labor

Another part of farming brought along a great many good folks, different as day and night. I am speaking of the ones you hired to help do different chores.

I suppose the first one that comes to mind would have to be Uncle Frank Anderson, who lived on Teaberry Road. He had a small farm where Gene Hughes now lives; seventy acres give a few or take a few. He had a threshing machine. I'm sure it was first class at one time. I remember it when it was about gone. He would start and work his way to your farm. Most always a farmer would go help one or two neighbors and in turn they would come help you. Uncle Frank was a preacher, had a church in Teaberry. They keep the old threshing box working by using baling wire. The old man called it the "bully thing" because it could be so difficult.

Mother cooked a week for a two-day threshing. Having the biggest barn around they threshed here, blowing the straw into the loft. We usually had two teams and wagon from here, one from Fullers, one from the Boone farm or sometimes Otts or Simmons or Grey Rock. It was a long haul; took three or four to keep the threshing box running. You needed three loaders in the field, pitching sheaves onto wagons, drivers, one man at the box to help unload, two sacking grain and the old man running the threshing box. It was run by an old Ferguson tractor and three men in the loft. Wheat was the big grain crop here, and also oats for horses.

Back to the dinner table on threshing day. We always had two tables full, ten at each, always with fried chicken, mashed potatoes, early peas, green beans, macaroni and cheese, gravy, rolls, and pies. The next day we'd have ham, biscuits, potatoes, stewed tomatoes, corn, caramel cake, pickles, strawberries, and fried apples.

A Mr. Longacre had a threshing box also. Same one as always cut our grain. Mr. Longacre and a man from Frankfort always pulled the wheat binder with three horses, and always cut around July fourth.

Blacksmiths were others who came to help. Most of the time they came to you, but in the spring when you were shoeing all the horses in new shoes you would take the horse to ride, and lead two or three. I think the first one I remember was Mr. Longacre, then Mr. Kipps. He was a good one, could make anything but he drank like a duck. He was also the only vet (self-taught) for years. A Mr. Charlton was a great one. He was one big man, he never let a horse get the upper hand, never. We took most all the horses to him. Buck Critzer was the last one to come. He was a hard drinker also.

Barton Ott came to pick corn in the winter. He also combined a few times. Now he is for sure one of the company's true characters. A true master of the machine shop, he could make you anything you needed or wanted. His wit was outstanding, truly a country man. It took food, food, food, to fill this man. I know 'cause I fed him! He needed a toddy to ward off the cold and chill. I suppose the sheep clippers and shearers were the others who came to help. I don't remember the early ones. I think Mr. Longacre from over the river in the Irish corner was the first I remember, then Cameron Wade, he drank a little, but he clipped for years. The last few times the Arbuckles clipped. We had twelve ewes and one buck of purebred Corriedale stock. The first year they clipped them it was a three-ring circus. It took Julian and Fil Arbuckle six hours to clip the buck, when done and the wool was gathered up, they had clipped fifty-six pounds from him.

Nowadays there just aren't as many folks around who do that sort of work. But these were fine folks, and I'm too glad to have known them.

Characters

From king apple we made many, many friends. The best kind of folks. I can recall some real old mountain characters. They would come the first of November each year. An old lady from Virginia and W. G. Robinson (Bill): they were both looking for cheap apples, and a lot of them!

First, the very old lady from the Allegheny Mountains of Virginia. Never knew anyone like her. She was a true mountain person. She always brought us something for colds that she made out of wild herbs. I don't recall her last name; we knew her as Georgia. She would have a big roll of clean sacks to put her apples in. In the center of the roll there was always a quart of moonshine! Her home was deep in the woods on the north end of the Little Allegheny Mountains. Georgia had an answer for everything. She smoked a pipe and chewed tobacco. She lived by the signs of the moon. She was one of those mountain people like you see on T.V. and movies once in a while.

The other person was a mountain man. He was Uncle Bill Robinson from out in Meadow Bluff district, somewhere in the hills off of Wilderness Road near Beaver Creek. His family was one that lived off the land, had four sons and two girls. None of them ever went to school. They were all grown when I came to know him. His wife died three years after we first knew him. He had no undertaker, just dug the grave and read over her himself. They raised everything they ate and used. Made their own sorghum, took grains to a mill in Fayette County for their flour, had bear and deer year-round to eat. The family lived totally to themselves. He was a well-learned man, self-taught. He believed he needed no help from anyone. They did buy shoes. Uncle Bill became a bitter man over the Second World War. He hid two sons from the draft, but both were taken to the Army later. One lost a leg; other one was killed. You never saw a man so bitter. The boy who lost his leg, when he got home, he got a car from the government, but the old man wouldn't let him bring it in sight of the house, he wouldn't ride in a car.

My brother Julian and I went out there one Saturday to bird hunt. We didn't get anything. They had seven killed and dressed for us, four quail and three grouse. The house was a log one, three rooms and a sleeping loft, big fireplace with everything on earth drying around it. The smell was bad. The cabin had been whitewashed, not clean, and smelled of tobacco and pipe smoke. They made moonshine but didn't sell any. It was for family only.

Uncle Bill doctored his own family and all his livestock. I think out there he was known as a pure devil. But he took up with us and we liked him. He would bring deer meat every time he came, also chewing tobacco for Julian that he had raised and cured. We never ate the deer, always gave it to one of the men.

To me it was a great day to have known both of these mountain folks and their family. The old gal never asked us to come up to her mountain. She wanted no one to know her whereabouts.

Butchering

Have just finished the hog butchering. We killed eight nice red hogs in early December and eight more here in January. Nowadays it's not butchering; we don't kill the hogs or dress the hogs. It's not butchering like done in the years gone by. No Sir! Not the same. There is no smoke in your eyes, no water freezing all over your clothes, no water splashing out of the scalding barrel over everyone, no ice-cold feet, no stench in your nose, the sickening smell of burning hairs, ashes, lime, no fingers so cold you think they will fall off.

Now when I think about all the butchering days on this farm, the ones I was set up out of the way, the ones I played around in the way, ones I begged to stay home from school to help and the many ones I have been a part of, I remember what fun I had, and how hard the men worked.

In years gone by butchering days were long hard ones: it was nothing for this farm to kill thirty-six corn fed red hogs. We tried all kinds of hogs. The Duroc, the Yorkshire, the Berkshire, Poland China were all too large, too fat. We wanted a lean bacon and ham hog, so we tried a red hog from Arkansas, a razor back wild type, then the red Tamworth. The farm started using them in 1923 and are still using Tamworths today. We always tried to down eight or nine before Thanksgiving. It was fine time; we were out of school and helped.

You could always tell butchering day was nearing: all the crew was busy getting ready. You could find old Jim the busiest man on the farm. Jim Early worked here over twenty-five years. He was a good man to me, a very old colored man. I always looked on him as a very, very old man, but he wasn't. He worked on the farm in his prime. A day or two before the set day you would find old Jim behind the woodshed setting on a chopping block, sharpening the gambling sticks and the spreaders;. he always made a few new ones. In the late fall when the first load of firewood was hauled in, a few good saplings for gambling sticks were on the load. Old Jim would lay them up to dry. He would also cut a basket of kindling for starting kettle fires, and he would clean and bed the chicken house a couple of days before, so nothing stood in his way getting ready for hog killing day.

You could find Dave Deaver cleaning the workshop, splitting kindling for the shop stoves, fetching a bucket or two of coal from the boiler room below, setting up worktables, getting down baskets. Later in the day you could hear the old foot powered grindstone humming away. In fact, if you stopped and listened you would be sure to hear Lynn Green singing along. It would be an old hymn. Lynn would be sharpening the axes, cleaver and all the butchering knives, plus a hog scraper or two. Lynn knew how to put a razor edge on the knives. Clarence Yates would be cleaning the old drag sled and the hay wagon. Uncle Jim and Charley Sullivan would be out at the spray house out in the orchard where the butchering was done; they would be setting the kettles. We used three black iron ones: one had to be fifty gallons, and small ones that held twenty-five gallons. They set the scraping platform by the scalding barrel, put up the pole, checked over a tripod pole or two. They had the buckets, baskets, hog scrapers and lots of pole wood for an all-day fire.

The butchering day started early, before daylight. Old Jim, who lived in Lewisburg and walked to work every day, walked up the orchard path to the spray house and would have the fire going by five-thirty. He would fill all three kettles. All the farm work had to be done before hog killing could start. Lynn came early and had all the milking done. Charley S. helped him. They milked eight cows, fed the calves, and cleaned the barn. Lynn took care of the milk and cream. Uncle Jim and Dave had taken care of and fed all the horses and cleaned the barn. They always harnessed one. It was Old Bill for many years. Henry Austin fed and cared for the sheep, no lambs this time of year. Clarence and Charley cared for all the hogs, sows, and pigs. The fat hogs for butchering were moved in the afternoon before from the outside fatting pen to a pen in the sow barn. This crew I am writing about was a crew with great pride, having worked together for a number of years.

Now looking back, some things they did were childish: everyone had his own scrapper, own rags, own knife, no one used the other fellows', no one. My Dad, Frank Cary, was a good butcher as was his brother, Uncle Jim. Their father, my grandfather, George Cary, had been a farmer and was a full-time butcher in later life. In fact, he was at one time a butcher in the southern army during the Civil War. Lynn was the sharpshooter of the crew; such pride he had in his one shot kill. Good man with a gun. He and Dave would turn two hogs out in the alleyway, run them down to the end where the outside door was. Lynn would down one, then one more. Uncle Jim and Clarence were outside with the horse and sled. Hogs were hooked and pulled outside; heads turned down hill. Uncle Jim would to the sticking. He always went in the left side into the jugular vein; he always said, "never a bad stick." Old Bill stood by, half asleep, the gun shots didn't bother him; he had been there too many times. Clarence would have the horse pull the sled that held the hogs and take them to the spray house. The horse had done this so much he could do it by himself. Hot water was ready; all were ready but Old Jim. When Old Bill got in sight with the two hogs, Old Jim would hurry on to our house, get his pint of coffee in his tin can cup, check the wood box, carry in a load or two of wood, get the coal in, take the ashes out, then go on to the chicken house to feed and water the chickens, then back to the house. He had to fetch up the sausage mill, lard press, sausage crocks, lard cans, salt back crock so my mother, Fannie Cary, could wash and scald so all would be ready the next morning. When Old Jim got back the first two hogs were scaled, scraped, tendons cut and pulled out, the gambling stick in place and the hogs were being hung on the pole.

The pole a was big one: one end was in a big fork of the apple tree and had a good brace at the other end, not a tripod, just a big set post. Sure didn't want that pole to come down! Henry would pour the last wash water over them; the loose hair and dirt was then scrapped off. Old Jim would be fixing the fires up, taking the scalding barrel water out, getting the kettle to reheat; all the kettles were refilled. Dad and Uncle Jim started the gutting. Old Jim helped; he took everything in the hog we didn't use. He always tied off the large intestine, then put all the guts in a basket or tub. Uncle Jim was a lefty, oh how he could use a knife! Insides of the hogs were cleaned and washed out several times with the spreader stick in place. They had to put the spreader stick in so the cool air could get inside. It would be cool air; always killed on cold days. By time the water was boiling, Lynn, Dave and Uncle Jim were back with two more. All the hearts and livers were hung on the brace post. When we were real young in the way kids are, Old Jim at noon time would get the hog bladder out and cleaned, then use a turkey feather to blow the bladder up, tie it off, and we had balloons. They always had six hogs hung by noon. Most everyone had a shared lunch on butchering day, homemade goodies to pass around. They ate only in between chores. All were hung by three and clean up started. Old Jim and Clarence would dump the kettles and put them up on a post barrel, then dumped buckets and scrappers were put in the spray house for another time. Lynn and Charley would get some liver and go in to pitch the corn silage down, feed cows and calves, and milk. Henry would go feed the sheep and all the surviving hogs. Clarence would keep an eye on the dressed hogs. Uncle Jim would go harness another horse, which would be a mule if he had his way. Uncle Jim took Old Bill with him to the barn. He and the other horse were hooked to the hay wagon driven back out to the spray house. All hogs were put on the wagon, and the livers and hearts were put in buckets and all taken to workshop. Here the hogs were hung on a pole inside. Old Jim would take the two lungs he had on the wagon into the chicken house and hang the lungs for the chickens to pick on. Then fed the chickens, watered them, and put on their night light, gathered the eggs, then got on to the house to put the eggs away, and get the wood and coal in for the night. By then the team had been put away and fed and all the work was done for the day.

Dad would always take Old Jim home with all of his hog goodies at butchering time. Uncle Jim rode along also, the rest walked home: Dave to Teaberry, Lynn, Clarence, and Henry to Lower Teaberry down around the Bethel Church. Charley went over the fence; he was our next-door neighbor. All had a mess of liver for the next day.

Old Jim was back early the next morning. He walked back; he was worn out as he had been up most all-night cleaning the intestines, stomach and all the inside parts. He came to the house, built the two fires, put the coffee on, did his chicken chores, got coal and wood in for the day, and took the ashes out. Lynn and Charley milked, which took more time this day. This was the day Lynn had to separate the milk, bottle milk and cream to be shipped to Charleston. Some of the cream was taken to our house for later churning. Uncle Jim would get the fire going in the shop stove, a very small fire; they couldn't have much fire. The rest fed and cared for the stock. Knives were re-sharpened, clean rags were torn up for all, and tables in right places. Baskets were ready with clean feed sacks in them. Everybody had a job and a good job they did. Dad trimmed the hams after Uncle Jim had blocked them out. He and Lynn did the shoulders and took care of all the bone meat. The rest cut sausage meat and cut the lard. When half of the hogs were cut up and a basket or two of sausage was cut, Dad and Old Jim would ready a clean table and dump a bushel of cut sausage, spread it out all over the table, then look it over. A good mix was the fat with the lean so it would be just right, not too fat. Dad would put half of the seasoning on the sausage and both of them would turn it on to the middle of the table, then spread it all out again and put on the other half of the seasoning, then mix, then rub it all together, and wipe up tabletop with meat to pick up all the seasoning mix. Old Jim was ready with the sausage mill set up and the hard job of grinding was ready to start. Others helped with the grinding as the day moved on. Lynn would poke up the fire and had a big iron skillet nearby. He made several sausage cakes; he would usually fry two skillets full. They always tried each batch. By then Old Jim had been to the house and mother had made a couple of pans of buttermilk biscuits for their lunch. She also made a cobbler. Lynn always had a homemade cake. He was a first class cook. He was a cook for three years in World War I. All the hogs were cut up by noon, and sausage making went on most of the afternoon.

Dad made all the mix for curing the meat. The first rub was made in the shop where it wasn't too cold, then all the hams, shoulders and bacon were taken to the meat house where the meat was laid on a bed of salt on a shelf. They always rubbed the skin side first, then all flesh was rubbed for some time until the cure went into the meat. More cure was put over the top. Ham took longer and more cure than the shoulders. Middlins (bacon) was last. They had to be weighed down or would curl. Dad used eight pounds of salt, three pounds of brown sugar, half a handful of red pepper, half a pound of black pepper, six ounces of saltpeter. This was enough cure for a hundred pounds of ham. The curing of fresh hams started with a cup of hot water mixed in the cure and then rub and rub they did. In five days, they would return and rub some more cure on, then again three days later.

My dad's sausage was the best. He liked it a little hot. For fifty pounds of good sausage, he liked it to be seventy-five percent lean and twenty-five percent fat. He'd use one pound of table salt, two large tablespoons of rubbed sage, four tablespoons of black pepper, two tablespoons red pepper. Sometimes he would pitch in a little ground cloves and nutmeg, about an ounce. While all this was going on Uncle Jim would bust the heads to get the brains out. They were taken to the house and put to soak overnight. The fat back and jowls were also taken to our house, rubbed in the sugar cure, placed in a big fifty-gallon stone crock jar, then a hot salt brine was poured over them. We had to cover them, and this cover was weighted down. The rest of the meat was taken care of: bones busted, ribs broken, feet cleaned once more by Old Jim and heads were cleaned and de-boned by Uncle Jim.

On the third morning Lynn and Old Jim would get on the lard making. This was done behind our woodshed. He always put up a wind break between the meat house and the woodshed. They used two big iron kettles, the cans were cleaned and set in a row on boards and were level. The lard press was also leveled. Cheesecloth was cut and ready. Lynn made the whitest lard he knew how; it had to have just the right fire at all times. He knew exactly when to put the next stick of wood to keep the fire burning even. As the lard cooked off Old Jim would dip, and Lynn pressed it out. As the cans were filled, they were set aside to cool.

The lids were then put on and they were carried to the basement, which had to be real cool. All the cracklings were saved. Mother made the best crackling corn bread. She always had Old Jim to chop them up fairly fine and she always used twice as much as most folks. All the men took a big mess home for their corn bread. Old Jim fed a bushel or two of the cracklings to the chickens. They loved the cracklings.

We always tried to let the hams lay for six weeks, shoulders four weeks, middlins twelve days. All were taken up, brushed very hard, then a wire was run through the hocks, a loop made on the other end and hung. We used a pig ring in the corner of the middlins, then they were hung. We had a real good meat house with a smoke furnace under the floor, but I don't remember them ever using this. Lynn was the smoker. He used an old iron pot in one end by the door, with hickory, apple, and corn cobs. He'd make a light fire and put a lid on the pot. He'd smoke for several days, but lightly smoked hams nine days. He had to tend the fire eight or nine times a day. The meat house is two rooms, hundreds of hangers in both rooms. The smoked meat was moved to upper room, dusted with borax, put in a brown paper sack, and rehung. The borax was put in top of the tied bags to keep flies out. The sausage was the next to be put away. Mother and Old Jim would fry out lots of sausage then place in crocks and top with grease. This was put in our cellar, some was put in cheesecloth bags and hung in the smoke house and smoked a day, then dipped in lard and rehung. Some was put away uncooked, just pressed in crocks and topped with lard. The loin was canned most of the time. Most of the time Uncle Jim and Aunt Fannie would work up the offal for us: livers, hearts, tongues, feet. Aunt Fannie made the best scrapple, liverwurst, and souse. She used liver, onions and pork broth and a little pork in her liverwurst. She had to keep an eye on Uncle Jim, he loved it red hot. He stayed home a couple of days to help. She made souse out of pigs' feet, a little head, two tablespoons salt, three-quarter cup vinegar, two cups pork broth, half a teaspoon black pepper, pinch of red pepper. It was good eating. Scrapple was corn meal and heads: four cups of broth, two pounds head meat cooked and pulled apart, two cups good white corn meal, one cup cold water, one teaspoon salt and lots of pepper. Mush was made out of broth water and meal cooked for half an hour and then meat was put in, cooked another twenty minutes, stirring all the time then poured out in a small cake pan or loaf pan and cooled, then you sliced, rolled in flour and fried brown. Best kind of eating. Longer your meat hung the better it was. Dad always wanted a three-hundred-and-sixty pound hog on foot, no larger, he didn't like the curing meat to go below 40 degrees and never over 60 degrees in the meat house when smoking.

Aunt Fannie always sent her makings to our home. She sent along a couple of stuffed cooked hearts, stuffed with bread dressing, which was very good. One morning we had a big mess of brains. We always scrambled eggs and cream, when half done put in brains, good eating!

These hog killings were repeated many, many times. Too many to count. We had sharp shooters, stickers, butchers. The crews were made up over the years by Curry, Hamricks, Legg, Brackman, Wall, Morris, Lynch, Hayes, Winalls, Sullivans, Austin, Yates, Green, Deaver, Williams, Miller, Arbaugh, Banton, Early, Cooley, Hughes, Hall, Zimmerman, Anderson, Blake families. We had lots of fathers, later had their sons and sons-in-law too.

Oh! That country ham, that white flicked country ham, red-eye gravy, and buttermilk biscuits!

Sawmill

Way back a few years ago West Virginia University had a forestry camp here in Greenbrier County, up on Anthony Creek. It was called Camp Wood. One of the faculty, by the name of J. P. Gull, built a portable sawmill. He wanted to show the farmers how to cut timber and then lumber it. He looked around for someone who would let him set up his mill and saw logs for them. He needed to set up the first time close by a wood. Tuscawilla had cut logs that winter; he found out and came by. I am thinking this was early May 1954. They moved in on a cool Friday morning, went on over the hill into the woods and picked out a spot for the mill. He had one student with him and three men, all working for the state of West Virginia. The mill was on a semi-trailer with its own diesel power; they were all weekend setting the mill and were going to start Monday morning. All the help we had to furnish were two men to stack the lumber. We were all there. J. P. was taking pictures of everything, but nothing went right as parts broke. They didn't get going until Wednesday. Everything worked well the rest of the week.

The next Monday, all the West Virginia University Forestry staff, the West Virginia Agriculture Commission, and federal commission folks were here. They were set up here about seven weeks. Lots of folks came to see the operations. It didn't take us long to find out that two men couldn't stack that lumber. We had old Sam Henson and his son, Everett, and sent for a couple of more. You never saw so much lumber. Most of it was pine and poplar, a few oak, and a hickory nut or two. They didn't use or move the mill all summer; they came back fair week and had more folks up to see the operations. The third week of September after a two-day rain on a very frosty morning, they had just got things going good when the man using the mill slipped and fell into the saw. He was killed on the spot. This was Paul Clendenin from up above Neola. He was a real tall man and had been a sawmill man all his life. He had a wooden leg and one eye; he had lost these in his work.

I was just opening the gate on the hill going into the pasture when I heard the mill stop and someone ran to the edge of the woods and yelled "Help, Help!" They saw me and asked me to get a doctor and to hurry. I was on a bay mare and rode to the brick house where I called Dr. Lemon. He was here in no time at all, and we had a truck ready for him to take him to the woods. The man was dead when the saw hit him. We called Wallace and Wallace Funeral Home. Paul Finely came for the body. He came back later and made a search for part of the man's head. He found part of his skull, but the eye and ear side were cut off. He found three other small parts.

Miss Elizabeth Knight was so upset and concerned, that we couldn't do enough for the family. They had nine kids, all living in two rooms. We took a load of food the next morning and money for whatever they needed. All the wife wanted was a box made out of the lumber he had just cut. The men here went to the woods and picked out wide pine boards. The state crew and J. P. Gull came, picked up the boards, and went to the forestry camp where they built a pine box for the old fellow. Wallace and Wallace didn't like this but that's the way it was. J. P. put a lining in the box, but the man's wife said no thank you, she just wanted a plain box. They buried him on their place. Three men from the farm went there to help dig the grave. Several of us went to the simplest funeral of all times. J. P. read a word or two out of the Bible; Paul Clendenin's kids carried the pine box to the grave. J. P. the next day picked out an oak board and made a headboard for the grave. He burned the words into the wood and then placed it on the grave.

The mill was pulled out in late September. We had a note or two from J. P. He moved the mill to Grant County. It hadn't worked out the way they had hoped it would.

One thing for sure though was that Tuscawilla had a huge pile of lumber, a big pile of sawdust and a large pile of bark. The lumber was used for the cattle barn on the east side of the farm. Five hog houses on runners were made and the apple room was lined up three feet high, so the grain could be stored. We used the sawdust around the trees for years. You didn't miss the trees that had been cut. I can't remember the board feet that were cut.

The year before the mill came here, all winter long we did nothing but cut timber. The trees had been marked ahead of time and we used two teams of skids for the logs and an old gray mule, and an old bay horse were used to pull the skids. Lord and behold they bought a big of team of Tennessee gray mules to pull the skids. Julian and Fil Arbuckle had gone to a horse sale in Tennessee and bought the mules. They were a fine team, prettiest dapple-gray things you ever saw. They matched to a tee!

Icehouse

The icehouse was built at the time of the barn, in 1920. It was three stories high with four sets of doors, real nice. About four feet inside the icehouse was filled with sawdust. The outer walls were covered tight so the saw dust could not get out. There was sawdust under the floor and sawdust overhead.

I can just remember putting ice in the icehouse. Back in the twenties, maybe the early thirties, they would gang up with all the teams around and go to Clifton on the coldest day of February and go to the deepest part of the river with ice saws and cut big blocks, three hundred pounds about. They would go back with wagons or a sled and used ice tongs to get the huge block of ice out of the river, then they'd put it in the icehouse. They used the ice all summer long.

I remember taking ice with Lynn Green, and Dave Deaver. Each of them, one at a time, with a mule and a cart, would haul ice over to the brick house, open the outside ice door, put it into the back of the old ice refrigerator in the hallway. I didn't ever see them haul ice from the river but have heard tales after tales about the ice gathering.

Poultry

Chickens here, turkeys there, ducks here, geese around, guineas, pigeons overhead, we had them all! We gave up all of these in 1956. We had a big hen house built around 1921. It had three large laying pens, two incubators, small rooms, a small hall, large baby chick room, sun porch, and each laying pen had a large outdoor run. Outside we had two brooder houses and pens for the chicks to live and keep warm. We set our own eggs in incubators and raised all the chickens for years. We had White Leghorn for layers, Cornish for bakers, and Rhode Island Reds for fryers.

We always had about three hundred laying hens, and always killed about five hundred fryers a year. They raised a hundred ducks, killed those, raised a hundred seventy-five turkeys, and killed those. Most of the turkeys were given as gifts at Christmastime to all the Knight families and kin and friends in Charleston. The geese were killed and eaten in the fall when Mr. Knight would come up to hunt. We also had a huge pen of pheasants. The pheasants were grain fed. A huge pen of quail were all put out in the fields. The quail would hear you and fly away.

One year we raised twelve kinds to show at the fair. We got day-old chicks each of Silver Laced Wyandotte, Plymouth Rock, Buff Cochins, Buff Orpington, White Rock, and Black Australorp. We liked the Black Australorp so much that we put them into laying and replaced the white hens. The duck house is near the chicken house, where we raised White Pekins. The turkey house and yard held the bronze turkeys. We also had Toulouse goslings and Chinese geese.

Brother Ed decided we'd give up poultry because it just wasn't cost effective for us anymore, cheap as store bought was. Tuscawilla's strengths were in the cattle and pigs, and from then we focused on those.

Cows

Where does Jersey cow cream come from? It comes from fine gentle, content, Jersey cows. Up until 1956, there was always a fine Jersey herd here on this farm. We tried to keep eight milking, and five to six replacer heifer and heifer calves. We had a purebred bull; bull calves were used for veal. Way back we didn't have to keep a bull; we took the cows to Red Oak Farm as they had their own purebred Jersey herd and bull. Their bull was named George Couch; he was large and black. Later they had a lighter black one with tan underneath: he was John Dickerson, and then a black one called Joe Dickerson. They came from the Dickerson herd in the Kanawha Valley and were purebred. Most of our cows were tan. In the early forties, we got some purebred ladies from Red House, Maryland. Six of the Collins' herd were Daisy Princess, Buttercup, Sugar Plum, Poppy, Grandy, and Tulip.

The first milkers I can remember were Harry Curry, Henry Austin, and Lynn Green. We had a maternity ward in the milk parlor calf pen. We had a carriage that ran on one track. It was a big dump tub to take the manure out of the barn and off to the manure pit, where it was dumped. Three kids could ride in it. You'd lower it by a chain and take it back up. What fun we had! It ran from the manure pit to the horse barn through swinging doors, to the dairy barn, back to the horse barn, then switch track up through other horse barn to other side of dairy barn. Oh! How much fun it was to put our city cousins in and turn it on. When it stopped, they looked out over the manure pit!

We started milking Holsteins in 1959 and 1960, with fifty head. The dairy was all modern. Lightning killed four cows in the woods one evening in an early summer storm. We all went to woods on foot, barefooted. I was there the next morning when they buried them. We raised two veal calves every year. The milk was shipped to the Knight family in Charleston twice a week, it was put on the train in Ronceverte. No rain, no snow, no sleet, or ice kept the milk from going. Bud Williams took the milk and eggs to the train, and sometimes my niece Kaaren would go with him.

The farm eventually became a commercial dairy. A milk truck came twice a week to empty the milk tank. At one time we were milking over one hundred head of cows.

Churning was done here on the farm twice every week, three time once in a while, but always twice. All the cows of the past I can remember were Jerseys, so lots of butter was made, the finest of butter from the finest of creams. I remember a large barrel on a rack with a hand crank; it must have been thirty gallons, but Old Jim is the first one that I can recall churning with the big and up and down dasher. It was a five-gallon churn, but he only put in four gallons of clabber cream.

No better butter was ever made than here! The milk was separated every morning at the barn. We kept the cream in the pantry in stoneware crocks. It was three days old when it was churned. Some days the butter would come right now. I have seen Old Jim churn three or four hours. He never gave up. Sometimes Harry Curry churned when Old Jim was too busy. Got a ten-gallon square can churn with a hand crank: you put the cream in, then the dasher, put the split lid on and away the churning would go. Law, sometimes soon, other times forever! Later on, we got an electric jar churn. We didn't have Jim and Harry in those days. But back to Old Jim's muscle power: he hand-cranked a wooden bucket around a metal container, with dasher and a sterilized lid while he sang.

Mother made most all of the butter. A woman named Birdie helped when she was here and well. Lots of other fine ladies made butter here: Nellie A., Lizzie Howard, Sally Austin, Estelle Williams, Laura Burns. Such pride my mother had in her butter. The butter would come out whitest yellow and was always put in a wooden butter tray. We used a paddle to work all the buttermilk and water out. The temperature of the cream was so important for churning. The temperature of the butter was very important to its outcome. You had to work butter very brisk, and it took water. We used to put a quarter teaspoon of fine salt to every pint of butter. I don't know how you know: good butter makers would just reach in the salt bowl and put the right amount of salt in their butter. We made square pound molds. Everything ever cooked on this hill from 1909 to 1958 was cooked with the best butter. Nine pounds of butter was mailed to the Knight families in Charleston every week. When there was surplus it was taken to the Woods store in Lewisburg, who had a waiting list for Tuscawilla butter.

Oh! that yellow speckled buttermilk, fine drinking.

We had other good things from that Jersey cream here on this hill. Folks were delighted to come for a meal or Christmas cheer and go on a picnic with a family. If you came for a meal, you had butter, cream, cottage cheese, milk, and homemade ice cream, the richest of all ice cream.

You just can't get butter that good at the store, and I wish people realized how different fresh, pure butter made from fresh cream is. It is the best kind for cooking or eating.

Fairs

For many, many years this family came down with an annual attack of late summer fever. It was surely a bad attack for some, others a light one, and one or two didn't come down with it at all. You didn't need a doctor to know what this fever was or what the cure was.

It was fair fever! My dad had the fever first each year and was the last to get over it. Just as soon as Old Man John Anderson drug out the whitewash buckets, ladders and brushes, and the fences, barns and the grounds started getting spruced up you knew the fever and the fair wasn't long coming. Dad loved the fair. He was a true horse person and when the first horses came on the fair grounds to train, he was there every evening, setting around the stables, swapping horse tales. He had great friends who came year after year. One of my very favorite fellows was Uncle Ben Hundley. He was from Ohio and brought three or four horses along each year. Uncle Ben was a poor horse man. He couldn't make ends meet. He tried. Uncle Ben named a bay mare, My Jane, after me, and as strange as it was, she was a good trotter. He won with her. But he pushed her too hard, over-doped her, and she was blind by age seven.

The old gent never came back after that. This farm never had a trotter or pacer.

This farm did show riding horses. We had at one fair three gaited horses, and a black dapple jumper by the name of Barronwood, a good horse. Another one was a tall slim sorrel horse called Highball; I might have mentioned him. He was an outstanding horse; showed in several shows each year. Another one, gray speckled in sorrel, named Rye, was too heavy but was a fine jumper. These are the ones I can remember. A fellow from Charleston by the name of Marshall Bond rode for Miss Elizabeth Knight; him I may have mentioned too. For years and years, I heard tales of Old Bob Tail Bess, Bessie Halstead, a few more tales of them going to the fair. The family rode these horses, the Knights, the Danas, and the Smiths.

As long as I can remember they always showed hogs, always the big red Tamworths. In the 1940s, Tuscawilla showed the new white Yorkshire and the reddish Durocs, and one year a couple of Berkshires. We kids were always around showing off, rubbing the hog's back, pulling his ears. About four years in a row the same man and lady brought an old Chester White sow from Virginia and won the call. We didn't take any more. It was bad for our ego, real bad. Having four purebred breeds on one small farm was too much, so we dropped all but the Tams. Way back then there were large hog shows. Folks from around the state showed. There was always a class for the biggest hog, so Tuscawilla kept the oldest boar over a year to fatten and show in this class. City folks couldn't get over them! This farm was more in the hog show than others. My Dad, his boss Mr. Ed Knight, and my brothers, Ed, and Julian, as well as Mr. Ed's son, Ed, ran the hog show for over thirty years. I would have to say it was a first-class show.

In the early fifties, Mr. Ed Knight was State of West Virginia Swine President, so those years, about eight in all, he and the State Swine

Association put on a feed at lunchtime the day the swine were shown. It fell to me to get the meal set up at the eating places. It was done in a big way: each year we fed more folks. Always had turkey, salad, hams: cooked from five to thirteen one year, plus potato salad, pimento cheese, pickles, apple pie and cheddar cheese, and pop, iced tea, and lemonade. It took a while to get ready. The afternoon before the lunch we made the turkey and pimento cheese sandwiches. Lots of great folks and friends helped. We would make twenty or more of each kind of sandwich. Each sandwich all of one kind were put in boxes, covered with wet towels, and placed into a cooler. The next day they were fresh, like you had just made them. The pies were made by a dear friend, Elsie Lee. We didn't mind the long day because of the fun we had.

Tuscawilla showed fruit for about the same years as swine: apples, peaches, cherries, pears, and plums. In the late thirties, we showed apricots. Mainly though we showed apples, always on the Monday morning of the fair, rain or shine, the apples were picked fresh. Lynn Green, with the help of Jim Early and Clarence Yates, would go all over the orchard picking a half a bushel of every kind that was in the show. Back to the orchard room, poured out with so much care, you then picked out five of the very best same size, same coloring, no blemishes. You had to take along three or four extras, if you would happen to drop one you needed another. Like the swine show there was a class for the biggest ones. Wolf River was the one for this call, one apple covered a paper plate. Everyone wanted them, and we got to sell them on Saturday. We never showed less than fifteen kinds of apples.

My sister Margaret later was made Superintendent of the Horticulture Department. She fancied up the place and put on a grand display of fruit. One year I recall it was so hot and dry in the Greenbrier Valley there was little on display at the fair. Mr. Ed let it be known that next year Tuscawilla would show in every department we could. I think that was the year before the war. Plans were made in September after the fair for the next fair. We decided what we would show, what classes we would show in, come fair the next year, we were ready.

First off, they showed a team of young draft horses. Charley Hughes cared for them and showed them as a team, then each by itself. Tea, a big good-looking horse, won her class in a big way. Coffee, a smaller mare than Tea, came in second. They were half brothers and sisters. Charley drove them in the cattle parade and used the farm wagon.

Next my brother Ed washed up and trimmed four pretty Jerseys. Two were so pretty, both heifers, other two were dry cows about to calf. These had to be registered purebreds and were, as I recall. One heifer won in three classes. We had the usual number of hogs. No one beat our Tamworth hogs. A full barrel wouldn't hold the hog ribbons over the years! We showed a very young herd of Shorthorn cattle, only six head I think, didn't fare too well with them. Likewise, the Jerseys, Miss Nettie Jackson's Jersey took the show that year. Valley View won the Shorthorn show. Tuscawilla placed in all classes but no big ribbon. We even took sheep, by golly, showed Hampshires. We also showed bronze turkeys: two hens and a tom. We showed white Peking ducks, a drake and two hens. Early that spring very early Mr. Ed Knight ordered twelve breeds of chickens to show, but we didn't show but six birds; you had to have a rooster and two hens in every class. We showed Silver Laced Wyandottes, Buff Orpington, Black Australorp, Buff Cochin, White Rocs, and a big game chicken. We raised them for years, the name I can't come up with, all won some kind of a ribbon.

One year we showed in all the produce classes. This was Lynn Green's pride and joy. He would come back on Saturday afternoon and pick what vegetables he could, and clean and finish on Monday along with the apple picking. We did well with both fruit and vegetables. After this fair we showed in all the classes for a few more years, but we dropped the sheep first, then the chickens, then ducks and turkeys, then vegetables. We haven't showed any of those for years. The State Fair here in Fairlea wasn't the first fair they would go to. They always went to Pocahontas County Fair at Marlinton. It was the week before the Greenbrier Valley Fair. Now this later became the West Virginia State Fair, in 1941. Brother Julian took the Shorthorn show herd for a full week. He used to get the cattle ready for showing, but with so much work to do on the farm, he had to get away to work with them, wash, trim and teach them to lead and stay tied up. That Pocahontas County Fair was a wild happy fun fair. The best kind of folks there, and they were happy to have you. Brother Julian had the best kind of friends up there. It was a hometown fair; only about two outsiders ever showed there.

Tuscawilla showed Shorthorn cattle at the Pocahontas County Fair, the Dunbar Fair, the Virginia State Fair, the Maryland State Fair, Roanoke Fair, North Carolina State Fair and the big show, the Southeastern in Atlanta, Georgia. When the Greenbrier Valley Fair was over, Julian started east to south. My brother Ed took the swine herd west to the Ohio State Fair and the Indiana State Fair.

One time at the Pocahontas County Fair, because we hadn't had the time to get the cattle in to train them, to wash them, to shape them, we took them up there on the river. We cooled them out, washed them in the Greenbrier River with cold water. It made a better coat, a much, much better coat, taught them to lead, taught them to stand, taught everything. Julian showed the cattle and Ed helped, but most times he showed the hogs and Julian had to travel. But we had friends. Every year you went to Pocahontas County Fair, everybody had a jug of moonshine up there, not just one or two but everybody up there had one. No outsiders came, nobody from out of state, just Tuscawilla and Greenbrier stock farms went up.

Ed took the hogs to the Minnesota State Fair, the Springfield Show and a couple three time to the Grand International way out in Chicago, Illinois, and it was always, always Thanksgiving week and the week after. They won top honors in the Tamworth show. He showed the grand champion boar and sow, and also a pen of three barrows and three gilts.

Back to the year we showed everything: we even showed dogs. There was a big dog show. Mr. Ed showed some working dogs, a pointer, a beagle and two fox hounds.

We really got started getting ready for the fair back in May. We always used wet days to get all the equipment ready. Show items were repaired and got a new coat of paint and new lettering. The pans, buckets, show boxes, fans, feed mixing box, cooker, show sticks, combs, and brushes were all checked to be in good working order. You had to clean show halters, oil show halters and lead straps. You had to make dozen or more rope halters. Mr. Ed was a great fair man, and you had to go first class or stay at home. He loved to go to the barn and set and tell about the history of the breed. He was a first-class mixer. He talked with all working folks, the owners, as well as the poor farmers. This outfit always took a gallon or so of the best moonshine around. It was in clean quart jars tucked in the middle of a hundred-pound sack of grain. It would take four or five sacks of grain for the moonshine. And if you were around when they loaded to go or unloaded when they arrived, you would see the big grain sacks getting loving care, just like a sack of fresh eggs! Mr. Ed Knight was a great sipper of the moonshine; he loved his drink.

There were always men who would help Julian with the cattle. Some that come to mind are George Lemon, Joe and David Williams, Red and Bill Rutherford. Julian and his helpers always did their sleeping in the barn. Fair time was long days, short nights, so you had to have two beds and bed clothes for the cattlemen, and one bed for my brother Ed, the farm manager.

Fairing was hard work, and in the long run rewarding for the farm. You made good contacts, often sold pigs, hogs and heifers and bulls to folks you met at the fairs all over. In your breed, the top winner had to furnish booze, the best to all who showed.

Julian Cary was one of the best cattle handlers around. He had a certain way with them. When he pulled into the unloading dock there was always three or four other breed handlers, all lending Julian a hand. They all wanted in return for him to help show their cattle. Not wash or get them ready, just lead them into the ring and show them. He won many a class by showmanship: he didn't have the best animals; he just showed the animal to the best a couple of years before they wore out. Even the Shorthorn folks wanted him to show. Whenever Tuscawilla didn't have anything in a class, he would show for others.

Julian and Ed made and were friends for many years with the finest of folks: Bill Laurence, a Herford man; Harvey Fishpaw, up in northern Virginia, Shorthorn all the way; the Wilsons; Harvey, Frank, Miss Mary, Edith, John, and Vint all from Valley View; Uncle Dick Johnson, an old bachelor from Monroe County; the Allens from Maryland; the Colemans from upper Virginia; Hank Davis, a friend for many years; Mrs. Faye Engles of the Hobby Horse Farm, Hot Springs, Virginia. These folks all showed about the same quality of cattle. One outfit from Ohio came sometimes to our fairs. He would beat us all; they were big time. Hog folks were fine folks too, many good friends among them: Glovers, Groves, Horns, Naltzes, McMillians, Masters, and Givens, all from Ohio and Indiana.

By darn I didn't tell you about Farmers' Day at the fair! It was the last Saturday. We had pony and farm horses, any kind of horse, mule races. Tuscawilla had four or five horses and three mules. Dad was up on some kind of a horse. I don't think he ever won. He would be pretty well loaded, a cigar and his hat turned back ready to go. One year had six riders and only five horses; the riders were Danas, Knights, Parliers. And, there was an old horse trader who had a very small exhibit on the fairgrounds, so Dad got the loan of a pretty good-looking bay horse, so they all rode. Guess who came in last? My old man. Gilbert Parker won one race. Ed Knight won one. Parlier's boys, Frank and Michael, came in first and second in another. Lots of grand old folks rode in these races, fun races. Charley Echols, Stuart Arbuckle, Julian Arbuckle, Fil Arbuckle, and Bud Rader would be in the mule race. The mule race was such that half were ridden in one direction while the other half were ridden in the opposite direction around the racetrack, and if you know mules, you would know that when they saw each other they all try to turn around and go the same way. Julian, Ed, Dad, Dave Deaver and Clayton Deaver even got Theady Mills drunk one Saturday and he rode. He was the only darkie I ever saw ride. Dave was the best hand at mule racing: he crammed his old dinner bucket full of rocks and he would shake it in the mule's ear. His pick was the old gray mule called Tazar; his half-brother was Arbury and he was bay, then we had a small gray one.

The friends we made at the fairs are some of the finest people, and we keep in touch best we can. Those were some great times, with plenty of good reward for our long and hard work.

The Cary-Hayes Families

Long Ago – Many, Many, Many Moons Ago

Note: This was written by Aunt Jane at my request. She's addressing me and my brother Frank. She believed that our families' history was important for the children to know, so they could know where they came from and from what kind of people.

You know, it's January 1995! Later than one thinks. I am going to try to put on paper a few hand-me-down facts of my family tree – both sides, Cary and Hayes. I am doing this for the ones still on the family tree. By the way, they are few and far between.

It's been a grand old tree. I like to think it's a big old apple tree. This old tree has been full of all the good things of life, though once or twice it had a bad spell or two – bent but not broken – that's for sure.

Here goes. First off, my dad – your Grandpa Frank – Frank Julian Cary was born March 28, 1884, north of Lewisburg. My mother Fannie – your Grandma Frances Lee Hayes – she was born November 26, 1891, also north of Lewisburg. They were married March 22, 1915, at Caldwell, West Virginia. Her Lee came from Robert E. Lee.

Ya'll are too young to know, but how one's memory can play tricks on you – Dad Frank as I remember him – he has been gone fifty-two years – was a strong but gentle person, in some ways a very tough guy. He was a very popular man, a polite humorous guy and was one of the biggest teasers of all times. He was a very devoted man to his family, his job, and his many friends. A very kind man who showed tears. He was not ashamed to shed tears as most men are. He was a man of solid judgment, and a better farmer than most. He was an outstanding horseman, and above all a good guy, a devoted father and also devoted to his fellow man in so many ways. We four kids, all four, could and did wrap him around our fingers any old time.

Now take mother, she wasn't like that at all. When she said "no" no it was! As I remember Mother Fannie, your grandmother, she was such a gentle lady who loved life and her family, her brothers and sisters. She showed her love in so many ways. She had so many loving friends and she gave love with true understanding to all. Mother hoped for nothing but the very best for all. She was a jolly gal. She was very contented to be the wife of Frank the farmer, and they were side by side for years. Mother loved all, never knew a stranger. She loved the little things in life we all walk past: sunrises, sunsets, the birds in the trees. Oh! How she loved the big full moon. Her favorite flower was the old-time hollyhock. She just loved them, and I keep some near the house now. She was a lady at all times. I never knew her not to be a lady. She loved her family; she was truly the drive behind this family. She ran the home and that's for sure! She was a proud person and would have liked to have been a private person but couldn't, too many needed her.

Mother and Dad didn't get to church much, chores were many and mostly up to our family. That's not to say they didn't believe in God: they did. Mother sent us to Sunday School – Margaret, Julian, and myself joined the Old Stone Church. Ed did also but he soon gave up the Old

Stone. He always said "too many blue blood Presbyterians looking down their noses," so he went to the Methodist Church in Lewisburg and sang in the choir. Mother joined the Old Stone in the summer of 1937. Mrs. Lonnie Dana stood up with her. I remember Mother crying coming back to her seat.

Mother and Dad were poor folks, real poor. Dad was married before. His first wife was Effie Ballard. She died in late 1914 and the baby died also. Dad became the farm manager here in 1916.

I don't know for sure if any of you all understand the way it was way back when the Carys first came to Tuscawilla Farm. E. W. Knight and family only were on the farm a few weeks. When the Knights were here in the brick house, Mother and Dad lived in the old cabin on the back road. It's where J. W. P. Knight (Whitley) lives today. Dad's mother lived there with Mother and Dad. When the Knights went back to Charleston, the Carys moved into the brick house. It was their total home. When Mr. E. W. Knight had to come up, Mother kept house for him and fed him. Margaret was born at the brick house. All this went on before time you know. I don't know when a person starts to have a good memory. I was sure I came on the earth remembering, knowing it all, not so, that's for sure. Anyway, I am going way, way back, so "furback" all this was handed down to me. So here we go again.

When the Carys lived at the brick house it was their home all the way, they had parties, kin folks dropping by, dances, picnics, big horseback rides, all the things that made for a happy home.

Tuscawilla started to have a face lift in 1916-17 and started on the new barns, this house, and a new water supply. Then the first World War came along, and most of the men went to war. Later the timber had been cut and a sawmill set up. It took a full three years to get it all up. After we got in this house, the brick was redone for the second time, then the big barn was finished. A couple of years later the cook's house was built, the five-car garage apartment, Aunt Ebbie's paint house, and the large chicken house. The meat house, the woodshed, the chicken house were all moved from the back yard of the brick house.

Mother raised us kids. She helped with so many chores here on the farm and she worked too hard for years, that is for sure. She saw that we had the cleanest milk at all times. She made the men wear white jackets. She then had that much more laundry to do. She helped with the meat house and the vegetable garden and then she raised all kinds of poultry. She had poultry all over the place. She had a big laying hen flock, loads of eggs. She set the old hen and raised chicks. Later she had an incubator, raised all the young chickens for eating and had a pen of turkeys, ducks, goslings and Guineas. Later on, we had a pen of ring neck pheasants and bob white quail.

Dad was farming. We had milk cows and a couple beef cows, lots of horses, some mules and some sheep. They got a water supply going and put water in the houses, so all the privies had to go.

When the family got moved to this house, the farm was going strong. It had a big crew, the fields were being cleared and plans for the orchard were in the making. The brick house had a crew: a cook, house boy, a gardener and a full-time chauffer. Dad ran the whole farm and Mother made butter and cottage cheese for all. She made sausage, saw to the lard making. Fat back was stored in the crock jars in our basement. She had lots to do with a very big vegetable garden. She cared for us and made the jam, jellies, pickles, and apple butter for the Knights also.

When they were in the brick house, that is the Carys, for winter, the Carys had a square dance. Ruth Hayes Smith, Agnes Hayes McClung, Mary Edgar Miller always told on each other. They sure told tales on each other and about all that came, the biggest finest horses, the rigs. The Tuckwiller brothers, Wilsons, Sullivans, Otts, Smiths, Brachmans, Hayes, the moonshine in the well house, how they danced all night, loads of them still there on Sunday morning. They would all wash up, put their ties on, and buggy off to church.

Aunt Bert and a few always told tales about the Ott grove where they had picnics in the woods behind the fairgrounds. They were all day affairs. Oh, how they told tales on the Hayes clan about going to the old-time church meeting in tents, most in and around Teaberry at one of the biggest ones ever held. The Hayes boys, mother's brothers, cut the ropes half-way through the meeting and the tent fell in on all! It took more than the Hayes brothers to cut all the ropes at once.

Another big outing for Mother and Dad was going once a year to the Silas Green Tent Show. It was an all-black show always held in Ronceverte. They both loved the show. It was a show of fun, jokes, dance, banjo playing – we kids got there once or twice, but I can only remember the bad smell.

Dad – Frank – was a first-class farmer, a good mixer wherever he would go. He and Mrs. W. E. Knight got along well. Dad ran the show: he hired, he fired, he bought, he sold, he swapped, and he traded. He was the boss. I found some of the old pay checks. Back in the 1919-1920 the amount was $55 for sixty hours a week.

Mr. W. E. Knight and Dad a had big thrill. Dad traded for a very young pair of white mares, draft horses. They were sisters, and he got them for a song. The owner sold them and went west. They were less than $50.00 each. He found a big reddish mule; they were bred and both men couldn't wait on the colts. One mare was named Hade. She had a grey male mule and that was Tazar. He was the pride and joy of both men. The other mare was named Maude. She had a reddish bay male colt they named him Arbury. They were born here on the farm and died here on the farm twenty-six or twenty-seven years later.

Your Grandma and Grandpa were both good neighbors, helped all. They helped the sick, helped with the funerals. I know the farm crew dug graves for years to help out. Dad always took his station wagon or sent it, and if it was a farm crew family, he would send the Knights' station wagon also.

I remember one bad winter with a real big snow, Dad and the whole crew were in the woodland cutting wood for everyone. The snow was deep. Mr. John Swift Dana, Sr., who lived on the hill, he was Mrs. Knight's brother, he and his family stayed for the first winter. He got off the hill. One daughter was teaching, one son was in Greenbrier Military School, and one son in grade school. On the way back home they got hung up in the lane and they pushed the car out of the snow drift and made it home. He then sat down in a chair and died.

Mr. Knight called from Charleston. Mother was a dear friend of Lonnie Dana – Mr. Dana Sr.'s wife – so she knew she was needed up there. She got together all the food stuff that was around and had her old faithful colored man, Jim Early, to walk in front and make a path up from our house to theirs. I went along and carried things but when we got there, I was sent home. Jim stayed and made paths. When Dad came out of the woods, I went and told him. He put a couple of the crew to feeding and milking and took the rest of us to Dana's house to drag the road clear. They opened the road and cleared it to the top of the hill. Mother stayed all night with Mrs. Dana. The crew didn't get finished until eight-thirty. Dad sent them home in the old farm truck. This was around 1936 or '37, I think.

I remember once all the Sullivans on the other side of the farm were sick. Mother took food and lots of good water and milk. She went every afternoon to see about them. I would walk with her to the fence, but I couldn't cross the fence. They all got better, but for Virginia and Charley's little girl. Most of the Sullivan men, in fact all, drank, so Mother went over one afternoon. They were going to clean up the place. She made a couple pumps; the cistern was dry. When they took the top off it was full of dead birds, rats. She made them clean the yards around the house and help in the house. The next day after they had milked here, she took Lynn Green and Jim Early over and they cleaned the cistern inch by inch. They lost the little girl. The last two or three days Mother was over there, when she came home, she undressed in the wood house and washed there, and her clothes were boiled in an iron tub the next morning.

By all means, by far the happiest outing for Mother and Dad once a year was the family reunion. The first one I can remember was also the last one they held at Blue Sulphur Springs in a big grove of trees. They were held there six or seven years. Then, in the Otts' Grove behind the fairgrounds, then one at Howards Creek at Caldwell, one just before you get to Blue Bend on Anthony Creek, then one at Blue Bend, and later Harts Run and a few here at the barn and in the woods. All the kin folks from far or near, lots of friends, all good folks tried for early starts and to stay all day. Truly there were some very strait-laced Hayes, but not too many. There was a much bigger hell raising bunch of Hayes. Oh, they were a fun, fun bunch. The men said their hellos in a hurry then to a table or a blanket on the ground for card games. There would be two or three real mad ladies, but drink and play they would. Aunt Zina from Rainelle would go to church first and would bring her preacher nine out of ten times, but he couldn't put the damper on the booze and cards. One group would be the happy go lucky bunch, the other group would be the chill group.

The fair was big in their lives. Dad started two or three weeks before, just as soon as the first horse showed up to train. All had to help get the work laid away early so all could be off. They ate a cold lunch and were off by noon. One of Mother's big things was she ate a fish sandwich and an orange drink about every day. We had to be home by four-thirty.

The milking had to be done, everything fed, re-cleaned up, then we ate a light supper and went back to the fair.

Back in those days the same show was put on for the six afternoons and the same show for the six nights. We went every time, you just couldn't miss one for anything. Mother tried for home right after the show was out, but Pop wouldn't be ready for an hour or two. He would meet up with the fellows, there was always a jug of moonshine. Mother would want us in the car by nine-thirty, but then she would send us out to look for Dad.

On Wednesday, Mother would pack a huge picnic basket, I mean a big, big, big one. We ate lunch and supper that day; all the kin folks came on Wednesday. We all ate together: Uncle Joe family, Uncle Will's family, Uncle Ben's family, Uncle Dick's family, Great Uncle John Hayes, Great Uncle Chris Hayes' family. All were good cooks, and lots of folks, you never saw so many folks. Men always had the moonshine jugs.

The fair was a real outing for Mother and Dad. The fair then wasn't like it is now. Back in their times the fair was where you saw your friends and family, some the last time was the last year's fair, so it was big.

Dad always rode in the farmer's races, mule races, etc. Ed and Julian did also before the farm started in the cattle. Most of the time the farm had so many wanting to ride, Dad had to pick up a horse or two from someone else. I remember one year Bob Follett from South Carolina had a rodeo fair week. For the Saturday races Dad got a couple of old horses for the farm to use. Dad rode one of them and when he was getting on the old horse it jumped back, caught Dad's pants on the fence and tore the seat out of his pants. I mean a bad hole. By the time the race was over, after a few stops at the White Lighting jug, he didn't care. They all went to the brick house for drinks and they took Dad. Aunt Ebbie went upstairs and made him a cover for his draws out of an old bed cover and took some old jewelry and made a big prize ribbon and they gave it to Dad, by then they were all high.

Your Grandpa and Grandma were big in the Shanghai Parade. He was always in the parade, mother only a few times, but she helped dress them all and sewed for them. Dad rode a big snow-white draft horse a few years, he was the parade marshal most years. Lynn Green and Dave Deaver always rode the mules. Claude McLaughlin and Mule Hayes drove some of the farm horses and an old car, always high as a Georgia Pine. Your Dad, Ed, was in for years. He was everything from a rabbit hunter, a cowboy, a big steer, a member of a black band. He and Charles McWhorter were in it for years and loved it. Julian, Shortie Arbuckle or Julian Arbuckle or all three were always in it. They were the Jessie James Gang, Civil War folks, or cowboys. Mule and Julian hooked a little grey mule in a buggy backwards. One year, what a time they had, had a bottle covered in a basket full of white lightning! Like father, like son, ha! One year much later Pete DeWeese and Lucille Brawley were up for the weekend. Alice Carr was out, all wanted to be in the parade. They talked Ed into driving the old, big apple truck. I slipped off and went to the brick house up to the attic, got into a lot of old clothes. I mean old that belonged to the Danas, Greenbrier Military School uniforms, etc. They dressed all over this house – took them to town in cars and then put them in the truck – the three old gals from Charleston: Mary Edgar Miller, Mur Hayes, Alice Carr and her sister Gladys, Alice Shields, Helen, Margaret, Sally, and many others were all dressed up and no place to go. Do you remember this year? You talk about a wild household. This one was the wild one, must have been twenty back to drink and eat and if one or two or more hadn't fallen by the wayside we never would have gotten them to bed, that's for sure. Fun folks.

A couple of big times were going to the circus, right good size ones, always up on the football field right in Uncle Joe's gate, and year after year around on Court Street where the houses are, where Soup and Curly Darnell lived, they also had other shows there, I can't recall the name.

I think I have said this somewhere else, here goes anyway. We four Carys, Margaret, Ed, Julian and myself, were all born here on Tuscawilla. Margaret over in the brick house, the rest here in this house. The same old lady helped with all four births, Miss Beckie Boone. Aunt Bert was here also. This is where Margaret got her nick name, "Beckie Boone." Miss Beckie came to help when we were sick.

When Margaret was about 6 or 7 years old, she got lost, and had the whole countryside looking for her. Oh, how Aunt Bert loved to tell this story! Alice Hayes and Black Hayes can recall this one. Uncle Bill and Aunt Mary Hayes lived at the end of the lane, and Margaret would take her pet lamb and walk down there to play. This day she didn't ask to go, so when she didn't show up for supper Dad went for her. She had been there and had started for home a couple of hours earlier. She couldn't be found. Someone told them they saw some men stop and pick her up, so a big hunt was on; they searched every inch of the farm buildings, barns, even the brick house. They couldn't find her or the lamb. They finally found the lamb asleep under a bush, and they called the law. When they saw the lamb, they were sure she was around, so they re-searched the barns and buildings and our house all over. Margaret was sound asleep behind a chair by the old-time radiator. They always said she knew they were looking for her but afraid she was going to get her behind busted for not telling them where she was going. She never did.

Ed was a little headstrong. He had a fast temper also. Dad put Ed and Julian to work at an early age. Ed spent lots of his time at Uncle Jim and Aunt Franny's. He was smoking by age of 6. Uncle Jim rolled his own, so he rolled Ed's. Mother all but locked him up, that is Ed. He was Aunt Fannies' pet, and I mean a big pet. He would get mad and take Old

Trigger, our dog, and run off to their house. They lived in the big house across the Teaberry Road. Grandmother Cary made this her home; in fact, they made their home with her. Ed was King Bee at Uncle Jim's and Aunt Fannie's. Ed was really the first one to get the name Teaberry.

We did all go to Sunday School, that is us kids. One Sunday at Sunday School, Ed had so much gum in his mouth, Miss Kate Moor asked him to get it out and he gave her a long spiel about how long he had wanted some teaberry gum and how hard he had to work to get that gum, no way could he give up his gum and waste his favorite teaberry gum. By the time we got down the street to Toots Grove's paint store where Dad was, the teaberry tale was on the street. They called Ed the "Teaberry Kid." He didn't like the kid part, so it was soon dropped. By first of the week the men at the Greenbrier Clothing Store started calling Dad Teaberry. From then on we were all called "Teaberry" or "tea." Most folks were sure the name came from the real teaberry. The real Teaberry is a lovely little bluff behind the Old Anderson overlooking the Greenbrier River right above Whitcomb on the old railroad. Teaberry was named for all the wild teaberry ground cover, it's full of little red berries.

Ed was in a bad dog fight, got his hand hurt. We were eating lunch one day when Mr. Knight's old dog, Kim, wandered down here. He was a springer; he was liver and white. He and Old Trigger, our dog, always got along. But a bad dog fight they did have. Ed ran out on the back, steps right in the middle of the fight. Ed got the whole back side of his left hand tore up bad, real bad. Dad had to take him to the hospital. It was about thirty days before he could milk or do much work. Old Trigger was our first, the one in the dog fight was the second Trigger. We were never without a dog or pups, I am sure. I can't name them, but I'll try for the heck of it. We had a light collie called Lady. She had pups: they were half bird dog. Old Spot, Uncle Jim's dog, was the father. We kept a pup and named it Lady Bug; she was by far the quietest dog of all. The next one was one of her pups, we called him Bark, that's all he did was bark. He was part bird dog also. We had yellow collies. We had a big one called Max, then we had Colonel which we got from Mrs. Prichard; he was pure bred. She called up, said "come pick out your pup." Sally and I went over and picked Colonel, a goodie. Then we had Major, then Sergeant Major. We also had a couple of bird dogs, Old Butch and Little Butch, the little black Scottie, Max, and now I have Mr. Red Eagle.

Ed got in trouble once again with a dog. One cold Sunday afternoon, Ed, Julian, and I were milking. We only had six cows milking then. Mother and Dad had gone to a funeral and were coming back when Ed saw an old dog out in the vegetable garden. He ran up here to the house and got his gun and shot the dog. He didn't kill the dog, only shot him up. The dog made it home, he belonged to Lynn Green's brother. They tracked the dog back up here to the farm. So, Ed had to walk down to Teaberry and tell the Greens about the dog and also pay for a new one. Dad wanted the dog killed, because he had been in the sheep, but he didn't want him to go home shot up.

We kids had some good fights; the boys never wanted to play with us girlies. They would try and run off from me, but I wouldn't give up. I hit Ed in the head, one day just before lunch, with a hammer. I was about seven, man alive I got grounded, couldn't go out of the yard for three weeks. He made me mad. Devils we were for sure.

Ed went to the barn one February morning, a real bad winter day, you could walk on top of the snow, about six inches, I think. I dressed and went later. We helped to water the sheep and cows. Our big sled was at the barn, so we got on up in the upper barnyard, we set Ed in front, and I was behind. We didn't need a push, just held up our feet and away we went, down over the large hill we flew. As we neared the bottom, Ed rolled off, and I hit a snow-covered stump. I met my match; I broke my right ankle in three places. Dave Deaver, who lived in Teaberry, had been walking to work for two hours one step at a time. He got me loaded on the sled and tried to pull me back to the barn and couldn't, so he had to carry me. Ed pulled the sled and carried Dave's dinner pail. I kept saying I was okay so when he got me over to the manure pit and I tried to stand up and walk but I couldn't.

Dad had caught up with us by then, and we were both in bad trouble. Pop got a man named Speed to help him carry me home. Old Dave was wore out. They got me upstairs to bed. Dad didn't think it was broken. Mothers' old helper Jim had just got to work. He had to walk all the way from Lewisburg. He went to the icehouse and cracked a couple of buckets of ice. They packed my ankle in ice. Two days no better, so Dad got Dr. Beard out and he said "broken." I was to keep the ice on it and meet Dr. Beard and Dr. Pole. The next morning at the Lewisburg Community House, they worked on my ankle. I spent the night there and went home the next day in a cast. I had bad scars for years, gone now. I was in the first grade and didn't get back to school until the next year. I was on crutches all summer. I had to repeat the first grade, but by second grade I was back playing football.

I was rotten so they say. I sure hated doctors, always cried when Dr. Beard came out or you had to go see him. When I was in the second grade, Dr. Arbuckle came to school to give school shots. I was in Mrs. Lipps class and you had to go upstairs for the shots. The class went but I wouldn't go, I had a crying fit. Mrs. Lipps sent for Margaret, and then asked Ed if he could get me upstairs. He told them to drop me on my head. Margaret sent for Helen Hayes Landers and I was drug upstairs and got my shot. I had a bad feeling about doctors, and I still have the same feeling. Before I tell a few more on me, I was spoiled rotten, a true pain in the neck, made the boys carry me everywhere, do everything for me, also got to go wherever Dad went.

In the third grade I was back on the football team and shooting marbles with the boys. I can hear Mrs. Lipps saying, "your mother can't keep you in clothes." How hard she worked to keep me dressed so nice for school. Having two brothers to kick me around I was a little rough. Lord when they were picking sides, I was always picked about third. I always wore a chicken feed dress. I never wore pants; no girls did, but I wore bib overalls at home. We all had a good grade school. Margaret would always tell on us. We all three played marbles. Ed was one of the best ones to play.

Dr. Richmond was our dentist. He and Dad hunted on and off together and drank together. In fact, Dr. Richmond drank most of the time anyway. In the spring some company gave him marbles to give his young clients, but he didn't. He would show up out here drinking and after a couple more drinks, Mother would cook supper for him. He would send the boys to his car for a big box of marbles. He was a jolly man. Ed got first pick. We all three had a big bag of marbles. We were always trading marbles.

Now Julian was a spoiled rotten kid. He got sick after supper one night, and by morning was in the hospital. He had appendicitis and was operated on the next morning. He was in the hospital eight days. Mother stayed three nights and by then he was the pet of all. You talk about a spoiled rotten kid, he was one. When he came home, we all had to look after him. We had to cater for months; he was one good-looking kid.

By the time Ed was in the sixth grade, the farm was a playground for all. All our cousins played here every weekend. In the eight-stall barn when the weather was bad, we played basketball in where the cows were milked. A big game was to see who could ride the manure carriage the fastest. It ran from the milk barn to the manure pit; it was a potbellied car and had a dumper on it. You could go pretty fast. Someone was always trying to pull the trigger on the side.

When my ankle was in the cast, I played a lot with Eddie Knight. He had a big case of allergies and Mrs. Knight wouldn't let him play at the barn, so we played up at his house. One big thing we did was we had a dry land turtle barn. We played under a big Pippin apple tree in front of the house on the hill. Big Mr. Knight had a bad skin problem, so Eddie wasn't let out in too much sun in case he got it too. Around the bottom of the old tree, we made little stalls with tie ups, each had a name, most had a date cut on the underside. We had to feed the turtles flies, berries, etc. We turned them loose every fall and started over the next summer. Most of them we found in the woods.

Eddie and I cooked supper every Thursday night up behind the house on the hill. He had a very small real stove; we had to get the wood and build the fire. We had to go to the vegetable garden and get what we wanted to eat. We cooked a little fish or chicken. Mrs. Knight's cook always fixed us brownies. Could she cook? Oh yes! Her name was Lumerna. Eddie's two cousins were there for a few weeks: Uncle Bill Paxton's daughters, Marlene and Martha. Abney Dana played also; he was Eddie's cousin. He was a sissy. He didn't like to play with the boys, ha! Geraldine Hayes Wetzel stayed with us in the summer, and she played also.

On hot summer weekends Dad would load us all up and take us to Blue Hole at Howard's Creek at Caldwell. We gave him fits; he was afraid of the water and sure kept an eye on us all. I remember one late Sunday afternoon, it clouded up. Dad said "load up" we got on Route 60 right in the middle of Caldwell and we were in the middle of a big cloud burst. We couldn't see, water ran in one door of the station wagon and out the other. We were devils. Dad all but killed us: we were fighting with wet towels, he all but made us walk home.

Fourth of July was big on the farm. Mr. Ed Knight put on one heck of a fireworks show. Mrs. Knight didn't want Eddie to help, so Ed and Julian helped Mr. Knight. I mean a big show: he made it bigger each year. Eddie did get to help later. All the Knight kin folks who lived up here for the summer would come. Some of their friends from Lewisburg, any of the help could come, lots of our kin. Mrs. Knight had punch. Mother, the cook, and Ethel Knight made cakes and cookies.

Mr. Knight took Eddie and Ed Cary camping a couple of times. Joe Cary would take them up the river above Renick and put their boat in the river. They would fish and camp on the Greenbrier and take three days to get to Ronceverte. Last time the boat came home with a hole in it, all three were burned real bad, real bad sun burn, so no more boating.

Also had big kick the cans in the barn. The barn would be full of all, the old and the young: Danas, Smiths, Prichards, Hardys, Prestons, Banks, Knights, Carys and kin folks. The old folks drank high balls, the young played kick the can. Mr. Ed Knight taught Eddie and your dad Ed how to shoot a gun and take care of one. He also taught Julian and me, too, but Julian didn't care for guns.

Right after Mother's death in May of 1941, the blasted war was on. That's when Aunt Ebbie's cousins from France and England came. They were third cousins; Katie Banks and her brother Noyes Smith helped to get them over here. Cousin Gertrude Dana married a French surgeon, Parlier, and her sister Mary Dana married J. F. Howard from England.

When the Germans got to the gate of Paris, Dr. Parlier turned coat and went over to the German Army. Mrs. Gertie grabbed up her children: Maddie, Frank, and Mike, without anything and went underground. They were hidden in France for three weeks, then they got into England, stayed there another two weeks. The Howards had some money, so they got them on a boat to the United States for safe keeping. The Howards sent their son Geoff and two sisters along. They went to

Dana's cousins in Tarrytown, New York. They came to Fairlea by the railroad, all but the two Dana girls, who stayed in New York.

I sure can recall the day they came. Aunt Ebbie gave them a walking tour of her beloved Tuscawilla. We were moving sheep when we met. Aunt Ebbie told the boys Dad was the boss and if he said "shape up" they had better shape up; he fell in love with all three boys on the spot. They were re-named Hillbilly Mike and Frank. Geoff, he called knot-head. He won them over right there on the spot, they melted right in front of us, they were his true pals.

Frank and Mike and Maddy studied at home here. Geoff went to town to school; he rode with Margaret. In fact, he was made Junior Class president. They drank wine three meals a day. We played lots of card games, ate together. They ate lunch here all the time. We did a lot of horseback riding. Geoff had a gal, Edith Houck, after him. She drove a pony cart and came about every day to visit. Geoff, with the help of Frank and Mike, built a tree seat in the very top of a big pine by the house. Every time she showed up, he would get up there with a book.

We walked to the river one summer three or four times. They would get home ahead of me by a half hour, but I did get home. I took them to the drive-in. We got pitched out of the theater a couple of times, so we didn't go back. In fact, all but got pitched out of the drive-in; they were cat birds, fun kids. They taught me to play bridge and we played a couple of times a week. They always talked to each other in French, and tried to teach me. Two or three times they had important evening guests, a French shipping line president, Ivory Soap man, a free French man. They ate and drank and we played cards; talked all night in French. We must have used ten bottles of wine.

Fred Hill from across the road (lived at Merry Hill now owned by the Hamiltons) rode over one morning. The word got out about Maddy's beauty and by the third summer they were married. He was the only son of the Arthur Hills. He had polio and was a little crippled in one leg; he came over every day. I think they had five kids. Fred died in 1986 or 1987. I saw Maddy in 1988 or 89. They all moved to Virginia after the war. Mike and Frank both got good jobs and married. Both have a family. Geoff went back to England. He married a real cute gal. They have all been back to the farm two or three times.

When Ed had a four-day pass in London, he called Geoff's mother and she picked him up, and he spent the four days on a big farm with her. Her oldest son and her husband were away doing army work. They were in timber and lumber. He was back to see her once again before he moved on into Germany. Mr. Howard and son were there this time.

Before Ed went overseas, he was home on a five-day pass with Charlie Lemon. The twins – Mike and Frank – wanted to have a birthday party for Aunt Ebbie's birthday, so the Rustic Inn was lined up. Away we went to a beer joint, Aunt Ebbie, Maddy, Margaret, Geoff, Frank, Mike, Ed, Julian, and Gertie Parlier. Maddy and the three boys were underage. Ed would buy the beer and pass it on to the boys; they had a couple of bottles of wine also and Aunt Ebbie had her bottle of gin. We were halfway finished eating our chicken dinner when a cab load of GIs came from Ashford Hospital (the Greenbrier). They picked at us and the next thing we were in the middle of a fight. Ed was high, he was taking on all. So, we had to get Ed out of there and home we came.

Mrs. Gertie (Parlier) fell off of Old Pete in the front field. She hadn't been on a horse for years. The boys had goaded her into riding. They were riding along when Old Rye started out in a run, they all three did. She was in the hospital for six weeks or more, then at the brick house with a nurse for six months. This was after Dad died in 1943.

Dad took them, that is the boys and us, to Roanoke, Virginia, to the big circus. Gas was rationed, and tires were rationed also. We had farm gas, but the tires were thin as paper. Away we went, and by the time we got to White Sulphur we had a flat tire and no spare. Dad caught a ride up to Forest Cary's garage. Forest jacked up someone else's car that they were working on, drove Dad back with two tires, and put both on the station wagon. We had to stay until the tent was down and all loaded up on the train; it was daylight when we got home.

The first year they were here we had a small fair, then no fair for four years. They were a great joy to Aunt Ebbie and all of us. Now it was another story with Mrs. Lucille, Mr. Ed and Ethel. They couldn't stand for Ebbie to keeping them, them being here on the place and the money she was using. It was her money and she loved doing this. Aunt Ebbie put Maddy's wedding on over at Warrenton Virginia Country Club, a big wedding. Oh, how Mrs. Lucille carried on over this, she tried to stop me seeing them. It was a bad time.

I was driving and took them somewhere every night, movies, parties, etc. Aunt Ebbie let them have about three big parties over at the brick house. The kids were all over that house, all in the gin and wine. They loved their parties. All the gals were after Geoff and Mike.

I remember Margaret going to 4-H camp four years in a row. These camps were held out around Rupert, a real swampy place. Dad always took the big truck and took them all. I remember being called one night; they had a bad rainstorm. So, Dad took two trucks and four men and went out to help them get everything out of the water. It was Margaret, Helen, Helen Goad, Margaret Fry, Martha Ford, Nancy Neal Coffman, Gay Tabscott, a Campbell, and a Laing. The same bunch of gals would go to Anthony Creek by the Greenbrier River. They had a big, big colored man, Bob Crawford. He was their cook and was there to keep away the boys.

I remember Margaret coming home from camp one day. It had been a wet week, and she was told to put the wet things on the line. She had been gone a week, so she had to go to town. She piled most everything on a radiator in her and my room. They heated and we had a fire. She got into bad trouble over this.

I don't recall Ed going to 4-H camp. He did go to Boy Scouts and football camp. They were both at Anthony Creek on the Greenbrier River. Coach Burr had a big football camp. They didn't have any money so all had to bring their own food. Robert Crawford was their cook. I remember Ed taking meat, potatoes, apples and all kinds of vegetables. I remember old Jim Early killing 14 chickens, and Helen Hayes Landers, Margaret, and Mrs. Burr fixed supper on Thursday night and took it up to them.

Julian went his freshman and sophomore years. They had money that year and the same cook. Julian didn't belong to scouts or 4-H. I belonged to 4-H and went to camp. One year Tri Hi Y had camp at Jacksons Mill for a long weekend. I didn't think much of that.

Ed got a car his last year of high school, a little Ford, a black one with two doors and a rumble seat. He was a biggie in that car. He had a girl in every class. I think he paid $66.00 for the car. He had to buy it a dollar down-a-dollar a week.

He stopped working on the farm that year. He was trying hard to get some money for college. So, Palmer Stacy and Spurgeon McQueen had been talking about a peach orchard on top of Muddy Creek Mountain. They gave Ed the job as long as Dad kept a helping hand there.

Ed had a four-man crew that cleared the land all summer and built fences. That fall Ed went to GMS. The next spring, they planted a big peach orchard and sprayed. He liked his job. He tried to drop out of GMS three times, he hated the place, but Dad and Margaret made him stay. The peach orchard never did get going. The freeze got it two years in a row and the deer ate it up. Ed then worked on the state road. He worked on Mike Landers' paint crew. They worked in four counties, so he was away most all summer.

Ed was in West Virginia University his sophomore year. There were seven boys up there. Richard Johnson was his roommate. Charlie McWhorter was up there also. He started off wheeling and dealing, but when the war broke out all got drafted and went in the service. All but Richard Johnson: Miss Mal and Miss Nela got him home. I don't remember when he went to the first camp. Anyway, Charlie Lemon was there, he was a captain. Ed didn't get in his company. They were in Virginia. Charlie came home every weekend and he got Ed out every weekend. Ed did the driving; Charlie drank his way home. They were moved to Kansas and were about ready to go overseas when Dad fell and broke his back. Charlie got Ed home in a day; he flew.

Dad fell down the stairs one cold morning. I was at the barn getting sausage to eat. He said he was okay and he walked back upstairs to bed. Dr Gunning came the next evening and put him in the hospital. He got along real good; in fact Ed was going back to camp on Monday. Dad died Sunday morning at six-thirty. A blood clot in his left knee.

He helped a sister move years before, and he and his brother-in-law were playing around with his dad's Civil War pistol. He got shot in the knee and they never did take the bullet out. I was up early that Sunday morning, getting a sporting company breakfast when the phone rang. I picked up the phone the same time Margaret did and the nurse who had been with Dad for a week had the sad news. We just loved her. Her hubby was over at Ashford General Hospital on duty. Martha Ranger, a cute gal. She was so hurt; she came by and spent the day with us. Ed stayed four more days. He got back to camp twenty hours before his unit was going overseas.

We had a bad time with the boys when Dad died. Geoff just couldn't take it. When Geoff found out someone sat with the body at night, he was going to stay with Dad. He started out but was talked into going home by midnight. He cried a full day and night; this was in December 1943. By then they were good pals of Uncle Joe Hayes and Uncle Jim Cary. Uncle Joe kept them in clothes and shoes. He always had the old back-room stuff full of all the things the GMS boys left, so he saw they got all they could use.

Mother and Dad always welcomed folks to their home with no fuss, no frills; everyone was just folks here. Mother could put one more plate on or four more. She always had more than enough for others. There is where I get fixing too much food. Dad was always bringing someone home for supper. Mother never said a word, just opened another jar and fed well. The cellar in this old house was always full to the top. There was a big garden here and she canned everything. I think she made more jellies, jams, etc. than anyone. She had so much fruit to work with. The cellar had jars of fat back, cans of lard, sausages in crocks.

About three times a year Dad had a big poker party and dinner. You think we set a big table now, you should have peeked at those tables: the biggest ham, the biggest turkey, both carved at the table. All the best dishes were used. A big blast, the table cleared and a long poker game, sometimes it would be daylight before they finished. Once they got snowed in until the men came and got the horses out to open the roads.

At family and friends' dinners, Mr. John McClung always carved the ham and turkey. Mrs. McClung made the cake. I remember Mr. McClung and Mr. Dana died the same week, two days apart. Mother cooked day and night for both homes. At Dad's poker parties, he always had homemade potato chips. Oh! So good so big. Mrs. Guy Dyche would make them. Dad would take 50 pounds of fresh lard, a sack of washed same size potatoes. Lord! She also made homemade candy and that she could do, she sure knew how.

Your grandfather was a first-class card player. He didn't have the money to be a good poker player, no way, but he held his own. He played with two doctors, two lawyers, one eye doctor, one real estate man, and one store owner. Now you take set back, he was a goodie at this game, couldn't be beat. After the apple crop was shipped off, he played set back every afternoon. All of us played set back, so did all the kin folks.

The boys played sports and got home some way. I studied more hours in the back of the pool hall – sitting in the shoeshine chair, if not there, at the Yarid's Sweet Shop. Mr. Alex always gave me a Coke. If not there up at Finely Arbuckle's Justice of the Peace office, so small, so smelly sometimes in back of the Greenbrier Clothing Company store.

Brother Ed sure hated the Army, and his dad sure hated the army. He tried everything to get him out and home but couldn't. Ed wouldn't have liked it; he would have been so unhappy if Dad had managed to keep him out. Dad didn't go to the First World War as he was a needed farmer. Ed got a battlefield commission. He made lieutenant while he was in Germany. He was hurt later in his fingers and arm, nothing bad. Ed came out of the war a very bitter man, I mean bitter, he wanted no part of the United States government, no part. He wouldn't take the job training. He could have been paid for three years. He came home on Saturday, drank too much, had a bad four or five days. He went to the barn, took the big apple truck over the hill for a wild ride, then on to town to Uncle Joe's. We got up there and got the keys from him. He was taken to Rosewood, was there for a couple hours, and then on to Aunt Fannie's and Uncle Jim's. Ed came out a lot harder than he went in. It took him some time to mellow out.

My, how one's mind wonders when you start recalling the past, it skips, hops, and jumps here, there, and yonder.

I was thinking how Ed had the mumps a long time before the rest of us. He was at Aunt Fannie's. He and Joe had them at the same time. We other three had them later. We all had the measles way back, and the three of us: Margaret, Ed and I had out tonsils taken out at the Community House (at the corner of Washington and Lafayette Streets) in Lewisburg. Miss Beckie Boone stayed with mother, and we stayed there two days and one night. I don't remember the boys ever being sick after the fifth grade. In fact, Margaret wasn't sick at all as I can recall. Now take me, that was my life! First grade, eighth grade, last year of high school, Lordy, Lordy! That's why Dad never had a spare dime, me!

Both boys played football and basketball. Ed not as much basketball as football. Both boys were good. Ed was his football team's captain both junior and senior years. He played guard and would never wear a helmet, never. The rules state that you must wear head gear so sometimes he was made to wear one. Before the game was very old, he would find some way to tear it up. Now if his mother showed up, he wore a helmet, yes, sir, he did. C. W. Burr was his coach for four years, and also Julian's for four years. Julian was on the team while he was in the eighth grade. He couldn't play but Burr needed boys to make up a second team so they could practice. Julian was a lineman for two years and fullback two years. He was team co-captain his junior year and captain his senior year. Julian was a good ball player; four to five colleges came to see him. Virginia Polytechnical Institute wanted him in a big way. North Carolina, Virginia, and North Carolina State asked him to come to check the schools over, but he wanted V.P.I. Mr. Alex Turner took him over there more than once.

Margaret was a good student, much better than the rest of us. She was in everything in school; on the paper staff four years, in plays, president of the senior class. She won the most improved student in her last year.

Ed was a wheeler and dealer. He was junior class president, president of student body his last year. He had a funny class, about thirty and only nine boys. So, they got their way, if not they wouldn't be in anything. All boys would hold out.

Julian was class president two years. He was in his class plays and was a big man at school.

Margaret went to Greenbrier College for Women for two years. She was one of the big town girls, she had lots of friends, about ten of the gals went on to West Virginia University.

Old Bud needed a medal, he drove Margaret to and from Morgantown for two years. Dad took her a couple of times. When Greenbrier College for Women was over, mother wasn't too well, but she dressed and went, all gussied up, hat and all. Margaret had a couple of grand gals as her best buddies, they had been in and out here. Mother was just wild about both and they about mother. Teddy was from New Jersey, she was some gal. When the good-byes came along, you never saw so many tears. Margaret was lucky, she had grand roomies at West Virginia University, Jane Nelson and Babe Moore. Jane came home with Margaret a couple of times.

Margaret tried for a teaching job in Lewisburg or Greenbrier County; nothing, so Mr. W. E. Knight got her on in Charleston. The day she was going down, they called up here a job was open at Rupert, and she took that job. She had a room and later moved in with Mary Edgar Hayes' mother and father, Uncle Cary and Aunt Bert. Margaret taught at Lewisburg High School thirteen years. She came the year after I left. She got her hands full. She started in Julian's last year. They gave her a hard time, she all but gave up. She did thirteen years and later went to the federal prison in Glen Ray, West Virginia. She was near the top on all the jobs.

Ed left dear old Lewisburg High School, went to Greenbrier Military School for one year. Lordy, Lordy! How he hated that school, he hated it to no end. He and Charley McWhorter were bad fellows.

They had more trouble at GMS that year from the town boys. They did everything that could be done to upset the school. One night the devils trapped a skunk. Someone knew how to trap one. They put it in a bag and took to the school, in Ed's car. Now, if you put the back legs off the ground, they don't put out the smell. They took it to study hall; it was small but full. Col. Joe Moore was the teacher; someway they got it out of the sack and the door closed, and then Ed and Charley looked in the window and about that time the skunk let go. It took the cleaners a week to get the uniforms skunkless. Ed was called up the next day. All the town boys who drove cars to school were called in and all cars had to be checked for skunk smell. Ed's passed. When school was over, Ed was out of uniform by the time he got to his car. He went to WVU, his second year along with Richard Johnson and Charley McWhorter. He was back to wheeling and dealing again. He ran for class president and he won, but the war ended that term.

Julian was all set to play football when the damn war started. He slipped off along with five classmates to enlist all trying to get into the naval air force; not a one of them passed. At that time, they were only taking the best. All got in later but Julian. When Dad died, Ed was in the army and Julian took over the farm. He was kept out of the army to farm. I didn't finish school, due to being in the hospital. I should have finished in 1942. By fall of 1943, I was helping Julian on the farm also. Dad died in December 1943.

We all four had happy school years wheeling and dealing, that's for sure. I was class president my freshman year, student body representative my sophomore year, junior class president, class president senior year, and Tri Hi Y Club president.

Margaret was about the only one who hit the books, we other three played around. They were all in their class plays, did a lot of running around at night. Ed did a lot of girl chasing after he got his Ford flipper, he had a gal in every class. Julian played the field until his junior year he fell madly in love with a cutest gal of all, Jeannie Powers. After they broke up, he didn't date anyone.

I recall Mr. Ed Knight got in the Shorthorn cattle at a meeting in Illinois. One time he found out there was a farm in Wisconsin looking for someone to help teach them how to show cattle. So, Mr. Ed made a deal with Dr. W. A. Taylor, he had a big ongoing Shorthorn farm. He was glad to have Julian come. Julian caught a bus in Lewisburg and rode the bus to Pardeeville, Wisconsin, in the dead of winter. He only had half enough clothes. It was cold out there. The farm was called Hickhouse or Hickouse, a real pretty place. Julian stayed with the manager and his wife. Folks by the name of Geff and Nan Perkins, older folks. Julian didn't do a daggone thing but haul manure. It was cold, cold, so cold you couldn't work the cattle. Julian learned a great deal about doing the cattle records. He came home with manure plaque the Perkins had made up for him. He liked both Dr. Taylor and the Perkins. They did show a few cattle that year around Wisconsin. Dr. Taylor sent Julian the first blue that Hickouse won along with a fifth of booze. Julian didn't even get to their town, cold, cold. He came home on the train. It was the coldest place he had ever been, cold, cold, cold. Dr. Taylor called Julian a couple of times and said, "if you ever want to relocate just call us, Hickouse would love to have you, anytime."

We had some bad winters. I mean good old fashion winters. I recall a few: 1959 and 1960 winter was a big bad one. We had around 96 inches of snow. It started around November and had six inches, December had a bunch of snow. January wasn't all that bad, maybe eleven inches. Now take February, that's when it got bad, forty inches. March had thirty inches, so much wind. A big bad storm or two.

I recall a few other things: Dad went to Detroit every three or four years. Always took someone with him. Soup Darnell if he could get him on his feet, if not Marion Hayes or a Lipps or a Cary from White Sulphur Springs. Mr. W. E. Knight sent Dad to buy two station wagons; they were cheaper in Michigan. If Dad couldn't get a ride with a dealer going up, he rode the bus. I remember six sets of Fords, two sets of Buicks, two of Pontiacs and four Jeeps. He always liked these trips. I recall one or two years we had show-and-tell supper parties, along with the Kyle Erwin's and the Purina Feed Co., always in the dairy and milking parlor. Food was light soup, or chili, always a good one. Showed how to feel calves, pigs, horses, all health plans for animals on the farm. The third year the Purina Feed Co. took several men to St. Louis, Missouri. Dad went as a hog farmer and a horseman, H. B. Moore for beef cattle, Guy B. Montgomery for dairy, the Wilson brothers beef cattle, and Edgar Dickson general farm animals. They went by train. Dad's first trip. When they got ready to eat the first night out all went to the diner, they accused Dad of thinking he was going to have to pay for it, so he had milk and crackers. He told them his "belly was upset", they gave him a big hard time on the trip and for a long time at home. He didn't care, he had a big time anyway.

I recall after Mother wasn't feeling up to par, Dad took her, and I rode along down on Coal River to the Jennings Brawley Camp. Lucille and lots of friends were there. Mother stayed a week, and she was homesick all week. Margaret and I rode with Dad to go for her.

I recall Mother telling me that in 1919 or thereabouts they took the horse and sleigh over Route 60 to sit with Dad's sister who was so ill. Mother told of the snow and ice up over the fence post – it held them all up. They crossed over where Holts Lane is now; it got 22 degrees below for three weeks.

I can remember Mother's house help. The first was Laura and Harry Curry, good colored folks who lived on Teaberry Road, on the curve where Herbert Montgomery now lives. They had a big family;

Harry was the house man for the brick house and helped Mother. Laura ironed for the brick house and also helped Mother. Jim Early came, he was here but first helped on the farm. Harry drank so much that they broke Jim into being a house boy. Then came Buddy and we had a little fellow from Ronceverte, Theady Miller. Mother bossed them all. Jim and Bud were here at the early times together.

Mother, your grandmother, loved the sounds and smells on the farm, from the outdoors to the inside. She loved to smell the meat house, the burning hickory, coal smoking fire, the gentle tang of the vegetable garden, smells of the corn fields, the wind falls of the apples that the cattle had crushed like wine. Hay, alfalfa, clover you could smell at night, the barn full of fresh oats, the manure keeps the barn full of smell. Mother loved the smell of the blooming apple orchard and when the buds started to open, she went every evening to smell and see the orchard, walked all over the place. It was a lovely sight, just suddenly over night the fifty acres with five thousand trees would burst into bloom, a pink and white sea, so many sweet fragrances.

Now I am writing about your grandmother Fannie. I had better tell you she was the champ buttermilk biscuit maker. Some cornbread maker too, in a big black iron skillet, crackling cornbread was a biggie on our winter table. I am sure her best dish of all was old chicken and dumplings. Once a week she would have an old hen ready, and when she got up, and after the old cook stove got up a good head of steam, she would take a lid off on the back burner, put the four- or five-gallon black iron pot on and in went an old fat hen. It stewed for hours slow and easy. Then a big batch of slick dumplings were cooked in, when the dumplings were done a big pan of gravy was made. So good. Her backbone and ribs were out of this world also. Folks would walk miles for her fried chicken and gravy.

Mother did a lot of drying of vegetables and fruit: leather breeches beans, the best kind of a bean, dried tons of apples for fried apple pies. One big treat was coming every year when the orchard came in bloom, mother had some three or four quarts of canned peaches put back and on the first Sunday of the bloom we would have our first homemade ice cream, peach no less, and a light wonderful cake with caramel icing.

I think we had a couple or three fires which Mother and Dad got so upset over. One May evening, Julian and I were told to sit out front. Mother was going to a movie and bingo. Dad took her. Oh, Mother was big on bingo. The Princess Theater started hosting. Mother won a couple big games. They didn't pay money, but they did in dishes, lamps, etc. Anyway before they came home, we had a big black thunderstorm to blow in, it was a biggie, lightning was so bad. They got home and we all went to bed. At this time, Margaret was in school at WVU, and Ed was in Summersville with Mike Landers and the state road paint crew, his first job for them. Along one-thirty lightning struck this house, it hit the chimney in Margaret's room, the one on the front by Mother and Dad's room. They were up, I was up, the chimney was closest to Julian; he had his head covered up, we could smell smoke, in fact the wallpaper was on fire. Dad got on the bed and put it out, there was still a strong smoke. We all carried all we could and walked to the dairy room. Then Dad said we could go over to the brick house. Rachel was over there; she was the cook. We went but couldn't get her to let us in, so Dad took us on to Uncle Joe's. Uncle Joe and Cary Moore came home with Julian and Dad. We called the fire department from Uncle Joe's. They were out, had five calls on the ridge from town to out here. We had no fire, only smut and dirt. What a mess. Had another bad one the next afternoon.

Another fire was in the fall. We all used wood and coal, all the ashes from our stove and the barn furnace were piled behind our wood house. T. Miller, a little old colored man, hauled them to the woods.

He wasn't supposed to touch them. We used them to unstick cars in the winter. Anyway, someone looked to the woods and it was on fire, so T. Miller stayed up all night. Dad hated fires, water, rivers, and lakes.

Another time, I was riding late one evening and came out of the woods and saw a big fire. It was the spray house, where we cooked the spray. Before I could get up there, the Fuller men saw it and came over and the fire truck was coming up the lane. Dang me, the old bay mare must have been slow. Ed was ill at the time, when that big, big fire burned Mr. and Mrs. Ed Knight's house up on the hill to the ground. Ed was so upset by this.

Back somewhere in this, I stated how much pride Dad had of the mules, horses, and the apple orchard. Well, he was so proud of his hams, in fact all the meat. Dad's grandfather was a butcher, and his father was a butcher. He was a butcher when he went into the army, the Civil War. Uncle Jim and Dad were both the best kind of butchers. So, he had a field day here on the farm, and butchering day was a big day.

Most of the early years they killed thirty or more hogs. Dad trimmed all the hams and bacons, the rest did the shoulder and jowls. Always just before Thanksgiving, Dad just knew the weather would be good, and Dad made all the sausages. Mother saw to all the fresh meat, lard, etc. We ate a lot of mush and cracklins. What we couldn't eat the chickens did. Mother and Aunt Fannie used every inch of the hogs. Dad loved the brains; we all ate brains. Dad put out a peppered ham no one could beat. We didn't butcher many beef until after we had lockers or freezers. We did butcher one a year, and if one had a broken leg or a young no breeder, we might do two. Dad would age the beef, he wanted it to hang twenty-six days. A big job to keep it clean and the flies out. Some of the smaller parts took fifteen days. Dad had a couple of old sayings: "oversee your workmen: if they be boys, separate them, for it is true, one boy is a boy, two boys are half a boy, but three boys are no boy at all."

I am sure I'll recall a few more words here and there.

I am sure of one big thing, Kaaren and Frank were badly cheated by not knowing their Grandpa Cary and their Grandma Cary. No two folks ever had more passion for life and other folks, no one. Their days' concern for others, their warmth, sharing and caring, you could almost say our life revolved around the kitchen table. It was always the warmest room, the old cook stove was great to sit by, and we ate, dressed, played, studied, and talked in the kitchen.

Back in the Depression Era, we never had a need so many did. Mother and Dad shared with so many. Your grandparents did so much for folks. They had so much here on the farm to share.

Your grandmother always did the biggest Monday wash: half ours and half the dairy's. The water was heated in the backyard and Old Jim carried it into the kitchen. She always did laundry in the kitchen. She always had three long lines of clothes, then all the barn towels were laid out on the ground. One Monday, Mother was taking a nap when a storm came up, so Jim pulled everything off the lines and put them on the kitchen table. He ran back out and jerked up the barn things and he started folding the cloths. Mother got up and, on her way downstairs she met an eight-foot black snake. Jim picked it up, he got it out of the house, didn't kill it. By next Monday there were two shorter lines up, no more ground.

Every Monday of the year everyone knew what supper would be, all you could eat. She would put the big black pot on in the stove hole and make a big pot of mush. Good stuff, sugar, and cow cream, you couldn't beat it. Tuesday morning you had fried mush, couldn't beat that down home eating.

Mother was the farm doctor. Every time anyone got hurt, it was Mother they came after. She had one or two bad days. One day at noon we were at the table, someone yelled. Jim Benton had been kicked, and here they came carrying him up under the apple tree, he had been kicked in the face. Jim and his family lived on the farm on Teaberry Road. He was unhooking a grey mare and a bay horse from a hay rake and for some reason the grey mare let go right in Jim's face. He needed to go to the hospital, but mother had to stop the bleeding, she packed his face in ice, got a clean shirt on him, and saw to a bed being made in the back of the station wagon and sent him on to the hospital.

Mother took her nap one day, got up and was walking to the vegetable garden. It was in the fall and she was going to pick some shellie beans. Walking by the hay loft she saw Lynn Green, who was the milker, laying in a heap on the floor. He had climbed to the top of the heap and was forking down hay for the dairy cows and had a bad fall. It was a long way down and she was sure he was dead. She found someone and they went for some of the other men. He was badly hurt, was out of it for five weeks. They put a steel plate in his head, he was in the hospital four months. They didn't have a car, no one drove in his family, so it was up to the farm to get his wife and son to the hospital and back. Mother went with his wife several times. He got over it in fine shape.

Your grandfather Frank was a cat bird. He took on a little too much hooch every now and then. If you saw him coming in the yard with his hat's front bill turned up, and half smoking and half chewing a cigar, he was loaded. Now if he was singing "Put On Your Old Grey Bonnet," or "Can The Circle Be Unbroken," Lordy, Lordy, he'd had a double load! And if he came up the back stairs to the old hymn "Rock of Ages," you had better get to bed. Mother would get up and sit with him and she would be a tough old gal the next day. In fact, he and the old colored man, Jim Early, kept her up in the fall along with Bud Williams when he and Charley Sullivan came. They would set an old hen, which was homemade ale, in the cellar. Everything had to be just right and what a smell! Mother was sure the preacher could smell it every time he came. To keep the help from drinking, it was bottled. They put it by the stairs. Unbeknownst to Mother, they put some under Ed's bed in the little bedroom off of the bathroom. It got hot and one night after midnight it started blowing up one at a time, there were twenty-four bottles. What a smell and mess. It left the house the next morning. No more homemade booze. Now they did set lots of cider at the barn and it would blow your hat off, all the way.

All and all, your family tree was a good old tree.

Apple

Recipes

Aunt Jane loved to cook, especially for family and friends. She learned to cook on a cast iron wood cook stove, the kind with the water reservoir on the side. I never did figure out how one could determine the oven temperature for baking on one of these. She learned to cook by helping her mother who was also an excellent cook. My grandmother's copy of the *Rumford Complete Cookbook* from 1918 is well-used and annotated with my grandmother's own techniques and recipes. Her annotations, though, frequently leave out the procedure, only listing the ingredients.

Aunt Jane had two secrets in cooking: one was butter. Having grown up on a farm and having a family in charge of the churning of butter you can see why. The other secret was cooking the food slowly. She would lower the temperature and cook the dish for a long time, especially any cut of meat. Most of her recipes were not written down: she kept them in her head from years of practice, and this is generally how I learned from her. Sometimes they would be included in her letters.

During hay season at the farm, she would prepare a hot meal for the men who were putting up the hay. When the hay is ready it has to be raked, bailed and stored in the barn before the rain came. It seems this always happened on one of the hottest days of the summer. Right after breakfast Aunt Jane would start working on a huge hot meal for the men putting up the hay. Quite often other farmers would come to help with this job and then the Tuscawilla men would help them. The meal would include roast, hot vegetables and salads and always cucumbers and onions in vinegar water with ice cubes to make it cold. I never knew how the men could eat so much food on such a hot day and then go out to work in the hay field. After lunch she would clean up the kitchen and start over with fixing another large hot meal for the men's evening meal.

Once when she was slowly frying slices of pork tenderloin, my husband happened to stop in. She was making dinner for the men at the Fairlea Exxon station just at the foot of the lane. Sam Rodgers, the manager, had been a family friend for many years; he and my uncle were best friends. Sam had a farm too and he often helped us put up hay. As she was stacking the cooked pieces of tenderloin on a large platter, Larry told her he would drive the meal to the gas station. Knowing how many pieces of her fried chicken he could eat she passed on this offer; she knew the platter would not be full by the time he arrived down the hill!

The following are a few of her recipes along with a few from my grandmother and one from a family friend. I have often thought her cooking for family and friends was her way of showing how much she cared for them.

Fannie Hayes Cary's Recipes

SOAP

5 lbs. warm grease

1 lb. can lye

1 qt. cold water

3 teaspoons borax

2 tablespoons sugar

1/2 cup cold water

1/4 cup ammonia

Dissolve lye in cold water, let cool, add fat slowly, stirring constantly. Mix other ingredients together, add to first mixture, stir the whole mixture until thick and light colored, pour into pans lined with cloths, cut, stack to dry, cut with wire or strong string, then stack to dry.

All the lard of the year before was made into soap or given to the crew to make their own soap.

COTTAGE CHEESE

For picnics, there's nothing better than homemade cottage cheese. If you were ever asked to bring a dish to a picnic, it was cottage cheese. So most all the time you could find a pan of milk on the back of the old wood stove. You used skimmed milk and you let it heat and clabber and when it was 75 percent firm; you cooked the curd hot for a while (half an hour). Then you cut the curd, and put in a whey porous sack to drain, then wash in cold water and drain, and work with your hands. It needs a big teaspoon of fine salt to one pound of cheese, cool then add pure cream. Will keep in refrigerator if covered.

CUSTARD

8 whole eggs

2 cups white sugar

1 teaspoon vanilla

2 cups whole milk

2 cups cream in a double boiler

Beat until all eggs are thick and creamy, add 1 teaspoon vanilla, make sure the water is boiling and never stop stirring. When a wooden spoon is coated remove the custard from the heat cool, use as soon as possible.

SNOW ICE CREAM

1 cup cream, add 1/2 cup sugar, 1/2 teaspoon vanilla, 1 egg yolk and snow. Mix until thick.

ICE CREAM

2 cups milk

4 cups heavy cream

1 cup sugar

4 tablespoons vanilla

¼ teaspoon salt

Scald the milk, add sugar and salt to milk, stir until all is dissolved, add cream, add vanilla. Cool, put in freezer. Makes 1/2 gallon.

EGGNOG

6 eggs, separated

1 pint cream

nutmeg

6 teaspoons sugar

12 teaspoons whiskey

Beat egg yolks, add sugar, some cream then the whiskey, then beaten egg whites, beat.

BLACKBERRY JAM CAKE

1 cup butter

3 eggs

1 cup sugar

1 cup blackberry jam or preserves

2 cups all-purpose flour

½ cup sour cream or buttermilk

1 teaspoon baking soda

1 teaspoon allspice

1 teaspoon nutmeg

2 teaspoons cinnamon

Preheat oven to 325 degrees (a moderate oven). Cream butter and sugar together, add eggs, beat until perfectly smooth. Add sour cream or buttermilk and flour alternately, beat 5 minutes, add blackberry jam or preserves, soda, allspice, nutmeg and cinnamon.

Pour batter into a greased cake pan or into greased cupcake pan. Bake in a moderate oven for 1 hour.

Cool for a few minutes and then take it out of the pan onto a serving plate. Ice with the following caramel icing recipe.

CARAMEL ICING

3 cups light brown sugar

1 cup cream (no half and half)

1 tablespoon butter (no margarine)

Add these ingredients in a pan over medium heat, stir well until mixed.Boil as rapidly as possible until a medium ball stage is reached (test in cold water). Never stir while boiling. If it should stick set it off the flame and beat in more butter. Once it reaches the ball stage remove from the heat and cool a few minutes.

Beat until it turns a light color, then ice the Blackberry Jam Cake.

Frank Cary's Recipes

HAM

Dad used eight pounds of salt, three pounds of brown sugar, half a handful of red pepper, half a pound of black pepper, six ounces of saltpeter. This was enough cure for 100 pounds of ham. The curing of fresh hams started with a cup of hot water mixed in the cure and then rub and rub they did. In five days, they would return and rub some more cure on, then again three days later.

SAUSAGE

He'd use one pound of table salt, two large tablespoons of rubbed sage, four tablespoons of black pepper, two tablespoons red pepper. Sometimes he would pitch in a little ground cloves and nutmeg, about an ounce.

OTHER PORK PRODUCTS

Mother made souse out of pigs' feet, a little head, two tablespoons salt, three-quarter cup vinegar, two cups pork broth, half a teaspoon black pepper, pinch of red pepper. It was good eating. Scrapple was corn meal and heads: four cups of broth, two pounds head meat cooked and pulled apart, two cups good white corn meal, one cup cold water, one teaspoon salt and lots of pepper. Mush was made out of broth water and meal cooked for half an hour and then meat was put in, cook another 20 minutes, stirring all the time then poured out in a small cake pan or loaf pan and cooled, then you sliced rolled in flour and fried brown.

Jane Cary's Recipes

AUNT JANE'S VEGETABLE SOUP

1 lb. ground beef

2 cups of chopped cabbage

1 large onion, sliced

1 stalk celery, chopped

3 carrots peeled and sliced

1 potato, scrubbed and chopped

1 Tablespoon parsley

1 6 oz can tomato paste

1 large can whole tomatoes, mashed

1 8 oz can beef broth and 2 cans of water

2 teaspoon salt

¼ teaspoon black pepper

1 ½ teaspoon sugar

1 package frozen peas

Brown ground beef in the soup pot, then the onions, then the potatoes, celery, tomatoes, tomato paste, cabbage, salt, pepper, then add beef broth and 2 cans of water, add frozen peas and other frozen vegetables if desired, (add sugar if the tomatoes are too sour) cook for an hour or a little more.

FRIED PORK TENDERLOIN

Slice a tenderloin into medium slices then pound them lightly, dip them in flour salt and pepper. Heat the skillet to medium and melt butter, place floured slices in the melted butter and slowly brown the slices turning frequently, cook until done to your likeness, add butter as needed.

A gravy can be made by adding milk to the drippings and stirring until thick.

BAKING A PORK ROAST OR TENDERLOIN

In order to keep a pork tenderloin from drying out when baking, place it in heavy foil, then place pieces of uncooked bacon on top, then add a can of crushed pineapple. Wrap the foil around the tenderloin and place in a shallow baking dish, bake in a slow oven for one to two hours deepening on the size of the tenderloin.

In order to keep a pork roast from drying out place it in heavy foil and spread a can of thawed undiluted orange juice on top, wrap in foil and place in a shallow baking dish, bake in a slow oven until done, again this depends on the size of the roast.

HOW TO COOK A SALT-CURED HAM

Take off the hide and fat first – cut the skin in 3 to 4 places and peel it off, then trim the fat so that there is about a quarter-inch of fat left. Soak the hock side of the ham in hot water for about 10 minutes to help get the pepper off, then drain the ham and brush off as much pepper as you can. With fat side up, wrap the ham lightly in foil so that it is airtight, then place in a 3 inch deep pan. Bake at 250 degrees for approximately 4 hours.

Check the ham for doneness by punching a sharp knife into the area around the hock, if the knife comes out with lots of juices, the ham is not done, continue to cook until the meat around the hock bone is movable.

When it is done, take it out of the foil, score the fat with brown sugar, mustard and whole cloves, the ham should be cooled before doing this.

Cut in thin slices for serving.

WILTED LETTUCE

Gather fresh leaf lettuce from the garden; rinse thoroughly, then drain. While the lettuce is draining, melt bacon grease in a cast iron skillet, or cook several slices of bacon in a cast iron skillet. Leaving the drippings in the skillet, remove the bacon to drain and add sugar and apple cider vinegar to taste to the skillet. Stir until there is a thick syrup. Add a little water and stir to combine. Pour this over the washed lettuce. Crumble the bacon slices over the top. Serve, and enjoy.

CUCUMBER AND ONIONS

Take cucumbers--fresh out of the garden are best--peel and then slice and arrange in a dish suitable for the table. Then slice an onion and place in the same dish. Add apple cider vinegar and a little water to taste, and a few ice cubes.

FRIED OYSTERS

One gallon of fresh oysters

Ritz Crackers crushed

Butter as needed

Egg wash

Drain oysters, squash the oysters flat by hand, dip in egg wash and crackers and cook slow in butter until brown, they should be soft to eat.

OYSTER STEW

One gallon of fresh oysters.

Mix half whole milk and half cream in a large soup pot, add salt to taste, add chopped potato and oysters. Cook until potatoes are soft.

Serve with oyster crackers.

FRIED CHICKEN, OR CHICKEN BIG JANE

My husband loves fried chicken. This is his favorite meal to request for his birthday or other special occasions. The first time Aunt Jane made it for him, he ate six pieces! Imagine an old-fashioned metal roasting pan full of chicken: that's how it was when she made this for the farm workers during hay making.

Whole chicken, cut up

milk

flour

salt and pepper to taste

butter

After cutting up the whole chicken, dip pieces in milk, then in flour mixed with salt and pepper. Melt butter in a cast iron skillet, place chicken and brown on all sides, adding butter as needed. Place browned chicken pieces on a rack in a roasting pan. Once all the pieces have been added, place roasting pan in the oven and water up to the bottom of the chicken.

Preheat oven to 325 degrees or lower depending on how long you want the chicken to cook. Roast/steam chicken for several hours, check occasionally to see if all the water has evaporated, if so, add a small amount. The chicken should be tender and falling from the bones.

Once done take out of oven, let it sit several minutes, remove chicken and keep warm until served.

Other recipes

EDIE SPRAGUE'S FROZEN DESSERT

3 eggs

½stick margarine

2/3 cup chopped nuts

3 bananas sliced

6 tablespoons lemon juice

6 tablespoons sugar

1 large can pineapple chunks, drained

2 jars maraschino cherries drained

1 ½ pints of whipping cream, whipped

In a pan beat eggs, add sugar, mix well and add lemon juice, cook over low heat until mixture thickens. Remove from heat, add margarine, stir, and then cool. After cool add nuts and the fruit, fold in the whipped cream, stir lightly. Pour into container and freeze until ready to serve. Cut into slices to serve.

This was a favorite Cary family recipe made for the Christmas Holidays.

THE BIG JANE SANDWICH

created by Billy Mike Landers

Slices of French bread
Slices of grilled ham
Slices of turkey
Slices of cooked bacon
Slices of Swiss Cheese
Mayonnaise for spreading on the bread slices
1 egg yolk beaten with cream

Spread mayonnaise on the inside of the slices of bread, add ham, turkey, bacon, and cheese. Dip the assembled sandwich in the beaten egg and cream, grill on both sides in a buttered pan. Serve with chips.

BACON CHEESE CASSEROLE: AN AUNT JANE SPECIAL

1 pound not too fat bacon

6 eggs

2 cups rich milk

2 slices of white bread cubed

1 teaspoon salt

1 teaspoon dry mustard

1 cup sharp grated cheese

Grease an 11½-by-7 inch dish. Cook bacon and drain well, beat eggs with milk and seasonings.

Put in a layer of bread cubes, then add bacon, crumbled, then add cheese, layer until all ingredients are used. Then pour egg milk mixture overall, place in the ice box overnight.Bake at 350 degrees for 45 minutes.

Serves 6-8

SALLY'S BISCUITS

4 cups all-purpose flour

5 teaspoons baking powder

¼ teaspoon of soda

1 ½ tablespoon sugar

1 teaspoon salt

Mix all the dry ingredients well and add 2 heaping tablespoons of Crisco. Now mix all together, make a well, add enough buttermilk to make a soft dough but pliable.

Dust the bread board with flour, knead lightly, real light, do not handle dough too much.

Roll out and cut small for a party, cut large for yourself. Pre-heat oven to 450 degrees. Bake until brown. Serve with butter.

MISS MAL JOHNSON'S FRUIT CAKE

Miss Mal Johnson and her sister, Nela, had a farm at the end of Rich Hollow Road near Hoffnagle Road. Their nephew Richard, who lived with them, was a friend and classmate of my dad's. The whole family had been friends with the Carys for many, many years.

Miss Nela, who was tall and thin, worked outside, taking care of the livestock, gathering the eggs, milking the cows and such. Miss Mal, who was shorter and heavier, was the housekeeper and cook. She was famous for her fruit cakes and in fact had a contract to make them for The Greenbrier Hotel. Who knows how many she would make each year! I can remember being in her large farm kitchen when it was full of fruit cake making supplies. The Carys always had one of her cakes at Christmas. She was so fond of my brother that she would make him his own small cake.

I am including her recipe as homemade fruit cakes seem to be a thing of the past. I hope that even if you are not particularly a fan of fruit cakes you will like hers.

1 ¼ lb. butter, creamed very lightly

5 boxes white raisins, washed and dried

1 ½ lb. sugar

2 lb. candied clear pineapple

2 lb. flour

½ lb. cherries candied

18 eggs, beaten separately until very light

¼ lb. citron, if desired

2 tbs. buttermilk

1 lb. blanched almonds

1/8 tsp. soda

½ shelled English walnuts

1 tbs. vanilla

½ lb. shelled pecans

1 can coconut

Add sugar to creamed butter. Add beaten eggs a little at a time. Mix two tablespoons buttermilk with soda and stir well. Combine the two mixtures. Add vanilla. Dredge fruit in the flour and pour batter over the dredged fruit. Bake in any size pan desired in a moderate oven for three hours. Makes 14 ½pounds.

Further Reading

Ruth Woods Dayton. Greenbrier Pioneers and Their Homes. Charleston: The West Virginia Publishing Company, 1942.

Ruth Woods Dayton. Lewisburg Landmarks. Charleston: Education Foundation Inc., 1957.

"National Register Information System". National Register of Historic Places. National Park Service. July 9, 2010.

C.E. Turley (January 1979). "National Register of Historic Places Inventory Nomination Form: Tuscawilla" (PDF). State of West Virginia, West Virginia Division of Culture and History, Historic Preservation. Retrieved 2011-08-03.

Lily Haxworth Wallace. The Rumford Complete Cookbook. Providence: Department of Home Economics of the Rumford Cookbook, 1918.

CPSIA information can be obtained
at www.ICGtesting.com
Printed in the USA
JSHW051115010422
24499JS00001B/3

9 781735 774718